FUNDAMENTALS OF QUANTUM MATERIALS

A Practical Guide to
Synthesis and Exploration

FUNDAMENTALS OF QUANTUM MATERIALS

A Practical Guide to Synthesis and Exploration

Editors

J. Paglione
University of Maryland, College Park, USA

N. P. Butch
National Institute of Standards and Technology, USA

E. E. Rodriguez
University of Maryland, College Park, USA

World Scientific

NEW JERSEY · LONDON · SINGAPORE · BEIJING · SHANGHAI · HONG KONG · TAIPEI · CHENNAI · TOKYO

Published by

World Scientific Publishing Co. Pte. Ltd.

5 Toh Tuck Link, Singapore 596224

USA office: 27 Warren Street, Suite 401-402, Hackensack, NJ 07601

UK office: 57 Shelton Street, Covent Garden, London WC2H 9HE

Library of Congress Cataloging-in-Publication Data
Names: Paglione, Johnpierre, editor. | Butch, Nicholas P., editor. |
    Rodriguez, Efrain (Efrain E.), editor.
Title: Fundamentals of quantum materials : a practical guide to synthesis and exploration /
    [edited by] Johnpierre Paglione, Nicholas P. Butch, Efrain Rodriguez.
Description: New Jersey : World Scientific, [2020] | Includes bibliographical references and index.
Identifiers: LCCN 2020029888 | ISBN 9789811219368 (hardcover) |
    ISBN 9789811219375 (ebook for institutions) | ISBN 9789811219382 (ebook for individuals)
Subjects: LCSH: Superconductors--Materials. | Semiconductors--Materials. |
    Crystal growth. | Intermetallic compounds. | Quantum theory.
Classification: LCC TK7872.S8 F86 2020 | DDC 620.11--dc23
LC record available at https://lccn.loc.gov/2020029888

British Library Cataloguing-in-Publication Data
A catalogue record for this book is available from the British Library.

The publisher thanks *American Chemical Society*, *John Wiley and Sons*, and *Elsevier* for granting permissions to the figures reprinted in this volume.

For any available supplementary material, please visit
https://www.worldscientific.com/worldscibooks/10.1142/11799#t=suppl

Typeset by Stallion Press
Email: enquiries@stallionpress.com

Preface

This book presents a comprehensive overview of modern techniques and strategies for the synthesis of quantum materials. It is intended as a starting point for graduate students, a reference for practicing researchers, and a supply of inspiration for quantum materials projects. The chapters in this book are a product of the lectures taught at the Fundamentals of Quantum Materials (FQM) school held annually each January at the University of Maryland in College Park.

The first FQM school addressed an unmet need amongst quantum materials researchers, particularly physicists. Despite the existence of a long tradition of sophisticated and creative materials synthesis, "sample growth" has historically been considered a dark art, and a sense of broader community was missing among practitioners of quantum materials synthesis. Now, the school brings together the next generation of growers to learn techniques and pointers directly from senior scientists. The enthusiasm from both students and teachers has been both gratifying and invigorating. After hosting four schools, we can confidently say that physicists, chemists, and materials scientists — experimentalists and theorists alike — all want to know how to make a good sample.

The point of this book is to bring the content of the FQM school to the quantum materials community. We would like to thank everyone who has been involved. We are indebted to all the lecturers and presenters who have participated in the FQM schools and workshops. We are grateful to the students, who have shaped FQM into what it currently is, and without whom FQM would not exist.

Finally, this book would not be possible without the commitment and expertise of the contributing authors, who have transformed their FQM lectures into an accessible manuscript. We hope readers will enjoy this essential guide and discover state-of-the-art techniques as they explore the Fundamentals of Quantum Materials.

Contents

Chapter 1

Introduction to the Synthesis of Quantum Materials: Some General Guidelines and A Few Tricks

Brian Sales

Materials Science and Technology Division Oak Ridge National Laboratory, Oak Ridge, TN 37831, USA
salesbc@ornl.gov

1.1. Introduction

This chapter is meant to provide a brief overview of how to prepare quantum materials, ideally as single crystals. The following chapters will discuss specific crystal growth methods for preparing quantum materials in more detail and with more examples. An obvious place to start is to ask exactly what is a quantum material? What are we trying to grow? This definition depends on whom you ask, but the definition that most people have adopted is the definition provided by the Department of Energy (DOE), a major funding source for the research. Quantum materials are defined as "solids with exotic physical properties, arising from the quantum mechanical properties of their constituent electrons, that have great scientific and/or technological potential." This rather broad definition evolved from a DOE workshop aimed at identifying the basic research needs for "Quantum Materials for Energy Relevant Technologies." The entire report is available [1]. A nice summary of the types of materials that have been grouped under the banner of quantum materials is given in a short article by Ball [2]. The list includes cuprate- and

1

iron-based superconductors, manganites, multiferroics, spin ices, graphene, topological insulators, spin liquids, Dirac semimetals and Weyl semimetals. These types of quantum materials can be roughly grouped as strongly correlated materials where magnetism is important, or topological materials, where the symmetry of the lattice provides protected electronic states on the surface or in the bulk of the crystal. As an interesting historical note, many of the ideas about the effects of topology and crystal symmetry on the nature of the electronic states near the Fermi energy (such as a Weyl semimetal) go back to Conyers Herring in 1937 [3], who also pointed out that similar effects should occur for phonons.

Since much of the field of quantum materials is driven by condensed matter physicists, it is perhaps not surprising that most of the materials of interest are inorganic compounds or alloys. The careful preparation of single crystals of quantum materials suitable for the investigation of the exotic physics is the focus of this book. In this chapter, we will discuss the "best practices" that apply to the synthesis of any material, a brief description of the various types of crystal growth techniques, and then a few examples of crystals grown by different methods in our laboratory at the Oak Ridge National Laboratory.

1.2. Basics: Best Practices

1.2.1. *Starting materials*

You have identified a potentially interesting new quantum material $A_x B_y C_z$ (where A, B, C are elements and x, y, z define the desired proportions) and you want to make it and study its properties. Where do you start? Almost all quantum materials are extremely sensitive to chemical impurities and defects. For example, the early attempts at preparing a topological insulator were unsuccessful because of the extreme requirements of composition and purity made the "insulating" bulk phase too electrically conducting to directly identify the topologically protected conducting surface layer. This is a problem in most semiconductors where the observed behavior is often due to unintentional doping or as Pauli noted the "dirt"

[4]. Strongly correlated materials, such as cuprate superconductors, have a similar sensitivity and small changes in composition can produce remarkable changes in properties. This means that your starting materials (A, B, C) should be as pure as possible. One common problem with purchasing chemicals from many vendors is that the purity stated on the bottle often excludes elements that may be present but were not specifically looked for by the vendor. For example, most vendors of rare-earth elements state a purity with respect to the presence of other metals, but nothing is said about oxygen, fluorine, carbon, chlorine or other elements. At low temperature small amounts of rare-earth oxide impurities can mask the intrinsic behavior of the quantum phase of interest. A more serious problem was encountered in our initial attempts to grow $RuCl_3$ single crystals, a potential quantum spin liquid. The starting material from the vendor was stated as being 99.99% $RuCl_3$, but upon careful analysis it only contained about 25% $RuCl_3$ with the remaining 75% consisting of $RuOCl_2$, Ru_2OCl_6, and Ru metal. In a couple of cases (over about 40 years of research), the bottle from the seller was completely mislabeled. In one case, zinc metal was labeled as antimony, and $CeH_{2.5}$ was labeled as Ce metal powder. The "Ce metal" powder was reacted with As at elevated temperatures to form CeAs in a sealed silica tube. While the CeAs did form, the reaction also resulted in a silica tube with about 10 bars of H_2 gas pressure. Upon carefully attempting to open the silica ampoule in a hood, the silica ampoule exploded from the pressure. Fortunately, no one was hurt, but this could have been serious and could have been avoided if we had initially x-rayed the "Ce powder" from the vendor. If possible, a good practice is to quickly examine a new batch of a starting material with an energy-dispersive X-ray measurement in an SEM and collect an X-ray powder diffraction pattern. In our lab, this total process takes about 30 minutes.

1.2.2. *Safety*

Some starting materials are either highly reactive, toxic or both, and it is extremely important to know how to handle each material safely. Although the Material Safety Data Sheets (MSDS) are sometimes helpful, under some circumstances all compounds or elements are dangerous (you can drown in water or suffocate in helium gas) and it is often difficult to extract the relative danger for a new compound from the MSDS information alone. We have found that a good supplemental reference [5] is "Hawley's Condensed Chemical Dictionary." This reference has a brief summary of the serious hazards and the Threshold Limit Values (TLV) for workroom exposure for a large number of inorganic and organic materials. The advantage of this reference is that you can quickly compare the relative hazard level of an unfamiliar compound to the hazard level of a familiar material that you know how to safely handle.

1.2.3. *Weighing*

To prepare a quantum material $A_xB_yC_z$, it is obvious that the pure starting materials must be carefully and accurately weighed in the correct ratios. What is perhaps not as obvious is the additional information that can be gained by weighing the reacted product at each synthesis step. For example, if you are making an intermetallic compound starting with elemental metals and the weight increases or decreases during a synthesis step, your sample likely oxidized, reacted with the container, or evaporated one of the elements. For more complex starting reactants such as carbonates or oxides, weighing can help you determine if your starting materials have additional water, or if the carbonate completely decomposes to the desired oxide. As an example, in the attempted preparation of Nd_2CuO_4, Nd_2O_3 was used directly from an old container of high purity Nd_2O_3. The standard synthesis protocol did not work because it was later found that over the years the fine Nd_2O_3 powder had adsorbed a lot of water from humid Tennessee air to the extent that 40% of the weight was water. Weighing at each synthesis step is particularly valuable if you are trying to make a quantum material for the first time. Chemical reactions do not always proceed as planned and tracking the weights at each stage can provide valuable clues that can help you modify the synthesis recipe. The famous French chemist Antoine Lavoisier (1743–1794) was able to make seminal contributions to our understanding of chemical reactions primarily because he constructed and used the most sensitive balance of that period [6]. An excellent three-part series from the Public Broadcasting Service (PBS) entitled "The Mystery of Matter" dramatizes the contributions of Lavoisier and his balance [6].

1.2.4. *Take extensive notes*

Growing a good crystal of a quantum material is difficult because there are a large number of variables and it is not always clear which variables are important. This is unlike commercial crystal growth of materials like silicon or germanium where all of the important variables have been identified through decades of careful research. For that reason, you should write down everything that you do during a crystal growth attempt: the batch number, purity and vendor for each starting material, how the starting materials were loaded into the crucibles and in what form (powders or chunks or prereacted material), which crucibles were used, weight at each reaction stage, which furnace was used, where the sample was placed in the furnace, programmed temperature profile for the furnace, humidity level in the room, etc. One reason for doing all this is that if you get a large high-quality crystal of a quantum material, you want to be able to repeat the process. An example of how important this can be will be discussed later in this chapter for the case of MnBi crystals.

1.2.5. *Carefully characterize what you grew before making extensive measurements*

William Hume-Rothery is considered to be one of the pioneers of the modern science of metallurgy. In the preface of his classic 1931 book on *The Metallic State*, Hume-Rothery comments on the differences between metallurgists and physicists [7]. To paraphrase his remarks "metallurgists make crappy measurements on beautiful samples, while physicists make beautiful measurements on crappy samples." Unfortunately, in too many cases this statement still rings true today. Most physicists want to get to the cool physics and exotic new quantum phenomena rather than carefully examine the crystal being measured. However, carefully characterizing your crystal can save you from an embarrassing mistake, which can have a significant effect on your career. In general, the more exotic or exciting the result, the more care must be taken in characterizing your crystal. Always measure more than one crystal. Large crystals grown by a Bridgman or floating zone technique are often not chemically homogeneous along the entire length. It is a good idea to sacrifice (cut into smaller pieces) at least one crystal to check for chemical homogeneity using an EDX or microprobe system, structural consistency with powder X-ray diffraction, and measure the physical property of interest (e.g., magnetic susceptibility or superconductivity) on different parts of the crystal. Crystals grown by flux techniques often have flux inclusions that are sometimes difficult to find. A specific example of this will be discussed in Section 1.4.

1.3. Crystal Growth Methods

The response of a crystal to external stimuli such as an electric or magnetic field depends on the orientation of the crystal with respect to the field direction. In general, the response can be characterized by a tensor or matrix of the appropriate rank [8]. Single crystals thus provide the best opportunity of understanding and tuning the response of a quantum material since more information is obtained as compared to a polycrystalline sample. Since many of the most interesting quantum materials are layered or quasi-two-dimensional (2D) with highly anisotropic properties, high quality single crystals are a must.

Most of the common crystal growth techniques can be divided into three categories depending on whether the crystals are grown from a vapor phase, grown by precipitation out of solution, or grown directly from the melt. Vapor-phase growth techniques include (1) vapor transport methods (*Haglund, Yan and Mandrus*) (2) molecular beam epitaxy (*Shabani*), (3) pulsed laser deposition and (4) combinatorial high throughput methods (*Yuan et al.*). Solution growth techniques include (5) metal flux growth (*Ribeiro and Canfield*) and (6) hydrothermal growth (*Wilfong, Zhou, and Rodriguez*). Melt growth methods include (7) floating zone synthesis (*Zoghlin and Wilson*) and (8) Czochralski and Bridgman growth (*Baumbach*), and the use of pressure (9) *Taufour*. The names in italics correspond to chapter authors for this book. A more complete list of synthesis methods and approaches can be found in [9] "Crystal Growth." Each crystal growth technique requires different experimental equipment, and for a particular quantum material deciding on a crystal growth method is one of the first questions to be answered. For most inorganic materials, the synthesis involves temperatures well above ambient, usually somewhere between 200 and 2000°C, and during some part of the synthesis the material has to be contained in a crucible made of a different material. As the reaction temperature gets higher, the number of suitable crucible materials gets smaller. Useful information about of suitability of standard crucibles for containing most elements in various temperature ranges is given in Table A.1 (adapted from Westbrook [10]).

1.4. Specific Examples of the Synthesis of Single Crystals of Quantum Materials

The examples discussed next are taken from recent or ongoing research at the Oak Ridge National Laboratory in the area of quantum materials. The examples are selected to illustrate different crystal growth methods as well the usefulness of some of the best practices discussed above.

1.4.1. *Vapor transport:* $RuCl_3$, CrI_3

In principle, vapor transport is one of the simplest and least expensive means of crystal growth. In some cases, the starting materials are simply sealed in a cleaned silica tube and placed in a furnace with the appropriate temperature profile and left alone for a few days to a month. This is basically how single crystals of $RuCl_3$ and CrI_3 shown in Figure 1.1 were grown. In other vapor transport growths, a small amount of an additional transport agent is used, such as a small excess of iodine or a compound such as $TeCl_4$ that

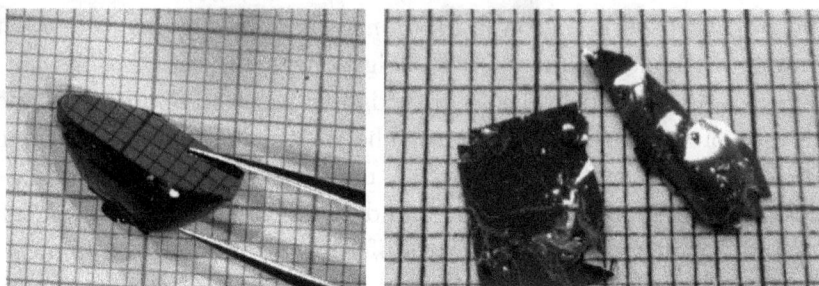

Figure 1.1.   Crystals of (left) RuCl$_3$ and (right) CrI$_3$ grown by vapor transport. RuCl$_3$ is a proximate quantum spin liquid [11, 12] while CrI$_3$ is a cleavable ferromagnet that remains ferromagnetic in the limit of one atomic layer [13–15].

decomposes to a gas at the appropriate temperature. The details of the thermodynamics of how this works are discussed in the chapter by Haglund, Yan and Mandrus, but the best practices discussed above are still helpful. If any of the starting materials tend to absorb water from the air, they must be carefully dried before sealing, otherwise the silica tube may explode on heating. The silica tube should be carefully cleaned, often with acid, and dried before loading the starting materials. The purity of the starting materials should be checked before starting. As noted above, the initial "high purity" RuCl$_3$ purchased from the vendor only contained about 25% RuCl$_3$ with the balance consisting of RuOCl$_2$, Ru$_2$OCl$_6$ and Ru metal.

1.4.2. Flux or solution growth: $K_2V_3O_8$, MnBi, $FeMnSi_{0.25}P_{0.75}$

The term "flux growth" is used in the literature to describe a type of solution growth, where the solution is usually not water, but the principle is the same. The flux is often a molten metal or a low melting point salt, or oxide. At high temperatures, all of the starting materials form a homogenous liquid. On cooling, a solid phase begins to precipitate out of solution. The trick with this method is to (1) make sure that the phase that you want precipitates first over an extended temperature range and (2) that the precipitation is slow enough to allow the formation of relatively large (mm–cm) single

Figure 1.2. (left) A 250 cc Pt crucible sealed inside a silica ampoule containing about 150 g of premelted KVO_3 "flux" plus 9 g of VO_2. After heating to 850°C followed by slow cooling to 400°C, $K_2V_3O_8$ crystals grew from the KVO_3 liquid. The KVO_3 flux was soluble in warm water, leaving black shiny $K_2V_3O_8$ crystals (right).

crystals. This is a very flexible technique that can be used to grow a lot of different materials. This technique also has the advantage that the processing temperatures are usually lower, which often leads to crystals with fewer defects. The drawbacks of this technique are (1) the crystals sometimes dissolve part of the "flux" into the lattice (2) pockets of flux are sometimes trapped within the crystals on cooling or (3) the flux is difficult to separate from the crystal. There are several excellent review articles on this technique [16–18] and the use of metal fluxes is discussed in the chapter by Ribeiro and Canfield.

$K_2V_3O_8$: In attempting to grow the potassium analog of the spin-Peierls compound [19] NaV_2O_5 (KV_2O_5) using a KVO_3 flux, large single crystals of $K_2V_3O_8$ grew instead (Figure 1.2). This sort of thing happens more frequently than is admitted in the literature when trying to grow crystals of a material for the first time. Rather than regard the unexpected phase as a failure, it should be considered a possible opportunity. This is one of the great advantages of making new materials, mother nature can often steer you toward something really interesting if you pay attention to what actually happens rather than what you wanted to happen. As it turned out, $K_2V_3O_8$ is a really interesting quasi-2D $S = \frac{1}{2}$ square-planar antiferromagnet where the magnetism and heat transport [20, 21] can be manipulated with relatively small magnetic fields. Although we first grew these crystals almost 20 years ago, they are still being investigated in the context of Skyrmion physics.

MnBi: MnBi is an interesting ferromagnet that has been known for over a century [22]. It has a Curie temperature of 630 K and was marketed in the 1950s as a permanent magnet "bismanol" with an energy product of 4.3 MGOe [23]. One interesting characteristic of MnBi is that the anisotropy energy increases with temperature from 100 K to 600 K. This is a highly desirable property for a permanent magnet and for the case of MnBi it was not fully understood why this happened. To better understand the intrinsic properties of MnBi, in 2014 we grew large (up to 1.9 g) isolated single crystals of MnBi out of an excess Bi "flux" using the published binary phase diagram. At the time we thought it was strange that no one previously had grown and studied MnBi crystals. After the first papers were published on the intrinsic properties of MnBi crystals [24, 25], a few more experiments were planned which required more MnBi crystals. Using the same recipe I developed 6–8 months previously, I found that only tiny 100 micron-sized crystals of MnBi grew. What changed! I tried different sources of the elements Mn and Bi, tried to further purify the Mn by arc-melting, tinkered with the heating and cooling profile — nothing worked. Finally, I went back to my original notebook and looked carefully for anything I wrote down about the growth of MnBi. Scrawled in the margin of one of the pages I noted that I had opened the furnace two or three times at about 1000°C and with the aid of long metal tongs swirled and shook the container with the Mn–Bi liquid. This, as it turned out, was an essential step in the ability to grow cm-sized crystals of MnBi, and it is a good example of why it is important to write everything down.

FeMnP$_{0.8}$Si$_{0.2}$: Single crystals (mm-sized) of this alloy are grown from a Sn flux. Alloys near this composition are of interest for magnetocaloric applications and for rich magnetism associated with a non-centrosymmetric crystal structure and two distinct magnetic sites [26]. We were excited to grow some of the first crystals in this interesting composition window and report on their properties. The initial resistivity measurements on a "single crystal" (Figure 1.3(a))

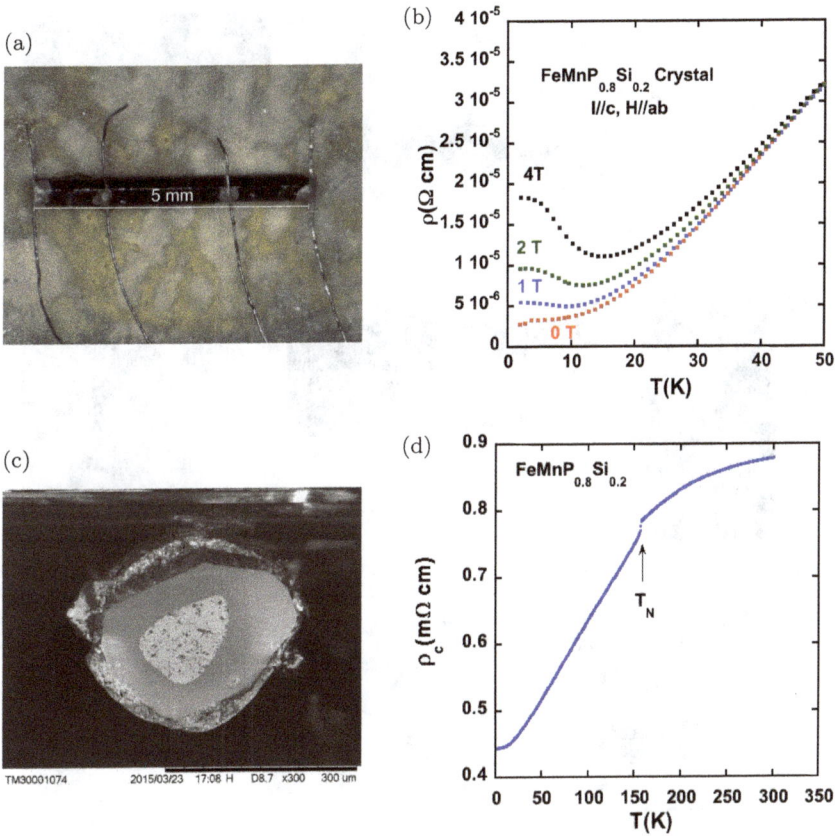

Figure 1.3. (a) "Single Crystal" of $FeMnP_{0.8}Si_{0.2}$. (b) Resistivity of "single crystal" shown in (a). (c) Scanning electron microscope image in backscattering mode of a cross-section of the crystal in (a). The bright region at the center of the image is tin metal that runs along the length of this crystal. (d) Resistivity of a crystal of $FeMnP_{0.8}Si_{0.2}$ after all of the tin was polished away.

of $FeMnP_{0.8}Si_{0.2}$ are shown in Figure 1.3(b). The resistivity is metallic with a residual resistivity of about $2\,\mu\Omega$-cm and shows substantial magnetoresistance at low temperatures. However, the magnitude of the resistivity is substantially less than reported for Fe_2P crystals [27], which did not seem plausible. For that reason, the crystal was examined more closely. Upon cutting the crystal open and polishing the end, we found that many of the "crystals" of $FeMnP_{0.8}Si_{0.2}$ had Sn cores running through the center of the sample (Figure 1.3(c)). Pure Sn metal has a huge magnetoresistance at low

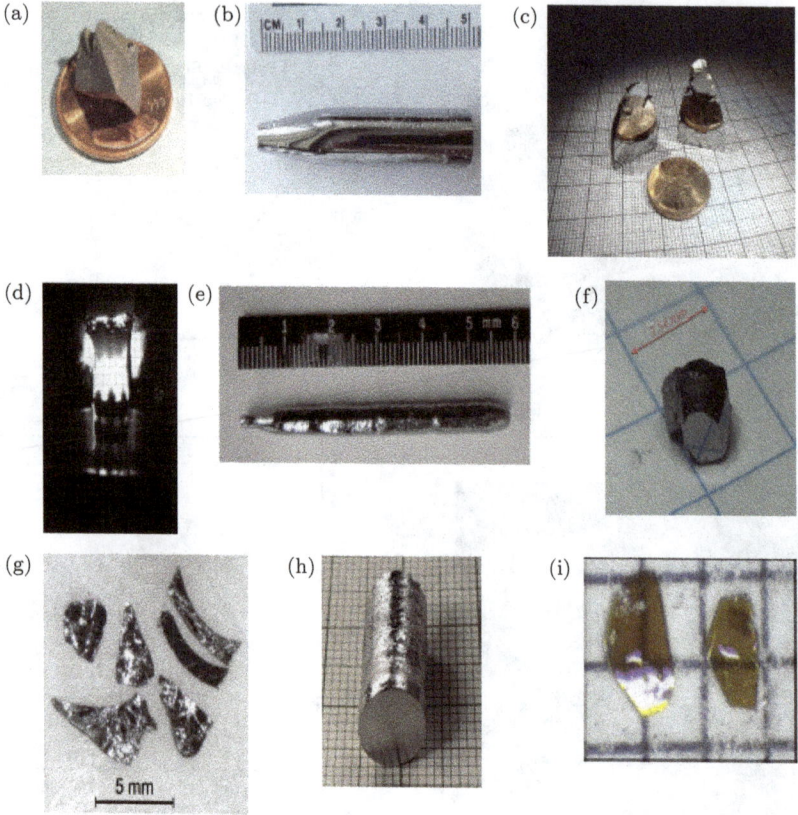

Figure 1.4.   Example of crystals of quantum materials grown at the Oak Ridge National Laboratory and technique used to grow them. (a) $CrSb_2$: Quasi-1D antiferromagnetic semiconductor with a giant thermopower [29, 30] — flux growth (b) $AgSbTe_2$: Spontaneous nanostructure — Bridgman [31]: (c) $Fe_{1.04}Te_{0.75}Se_{0.25}$: Iron-based superconductor — Bridgman [32] (d) $BaFe_{12}O_{19}$: quantum para-electric — high-pressure floating zone [33] (e) $Fe_{0.96}Ir_{0.04}Si$: resonant phonon scattering [34] — floating zone (f) Mo_3Sb_7: superconductor with spin dimers — horizontal flux growth [35, 36] (g) $MoCl_3$: magnetic above 580 K — vapor transport [37] (h) NiCoCr: medium entropy alloy with itinerant quantum critical point — floating zone [38] (i) $Cu_{1-x}In_{1+x/3}P_2S_6$: tunable cleavable ferroelectric — vapor transport [39].

temperatures [28]. If the Sn core was polished away, the remaining single crystal of just $FeMnP_{0.8}Si_{0.2}$ had a residual resistivity of 440 $\mu\Omega$-cm and virtually zero magnetoresistance [28]. This is why it is important to carefully examine and characterize what you are measuring before thinking about modeling and explaining the physics. Other crystals of quantum materials grown recently in our laboratory along with the science drivers and growth methods are shown in Figure 1.4.

The discovery and investigation of new quantum materials is both fun and scientifically challenging. This introduction hopefully conveyed some of that excitement while pointing out some of the best practices that will help ensure the success of *your* research.

Acknowledgments

I would like to thank Andrew May, Michael McGuire and Jiaqiang Yan for reading and helping improve the Chapter. This work was supported by the Department of Energy, Office of Science, Basic Energy Sciences, Materials Sciences and Engineering Division.

Appendix

Table A.1.   Atmospheres and containers for the melting of metals and alloys.

Class	Metal	Refractory for crucible	Protective atmosphere or flux
Alkali metals	Li Na K Rb Cs	Pyrex or silica at low temperatures; steel at higher temperatures	Argon preferable; all except Li can be melted under heavy oil or paraffin
Alkaline earth metals +Al	Be	BeO or ThO_2	Argon
	Mg	Al_2O_3, MgO, graphite or steel	Argon or sulfur vapor; under KCl, $CaCl_2$, CaF_2 flux; avoid nitrogen
	Ca Sr Ba	Steel, graphite for Ba	Argon
	Al	Al_2O_3, MgO, BeO, CeS graphite; avoid siliceous material	In argon under charcoal, can be melted in air without flux; N_2 or H_2 in graphite

(Continued)

Table A.1.   (*Continued*)

Class	Metal	Refractory for crucible	Protective atmosphere or flux
Refractory metals	Ti Zr Hf	Mo_3Al for Ti; arc melting on copper mold	In argon or vacuum
	V	ThO_2	In argon or vacuum
	Nb Ta	See Ti, Zr, Hf; for Nb: ThO_2 or BeO	Hydrogen
	Cr	ThO_2, ThO_2 lined Al_2O_3, BeO; reacts slightly with Al_2O_3	Hydrogen or argon
	Mo W	Arc melting on water-cooled copper mold	Hydrogen or argon
Th + U	Th	ThO_2 results in some ThO_2 pick up; does not react with CaO or BaO	In argon or vacuum
	U	BeO} ThO_2} up to 1500°C, CeS to 1900°C MgO}	In argon or vacuum
Mn + Re	Mn	Al_2O_3, spinel	Hydrogen or argon; too volatile for vacuum
	Re	Arc melting on water-cooled copper mold	In argon, vacuum, or nitrogen
Iron group	Fe	Al_2O_3, BeO, CaO, ZrO_2, ThO_2	Hydrogen
	Co	Al_2O_3, ZrO_2	In hydrogen, argon, or vacuum
	Ni	Al_2O_3, BeO, ZrO_2, ThO_2	In argon or vacuum, H_2 or He for BeO, ZrO_2, ThO_2
Group VIII precious metals	Ru Os	ThO_2 or ZrO_2	In nitrogen, hydrogen, or vacuum
	Rh Ir	ThO_2 or ZrO_2	Nitrogen or argon
	Pd Pt	ZrS, ZrO_2, Al_2O_3, CaO	Carbon monoxide, nitrogen, or argon

Table A.1. (*Continued*)

Class	Metal	Refractory for crucible	Protective atmosphere or flux
Copper, silver, and gold	Cu Ag Au	Graphite, Al_2O_3, MgO, spinel	Under charcoal, nitrogen, argon, or carbon monoxide; hydrogen in graphite
Zinc, cadmium, and mercury	Zn Cd	Pyrex at low temperatures; Al_2O_3, graphite, mullite	Under charcoal or halide fluxes, argon
	Hg	Pyrex	In argon or vacuum
Group III	Ga	Pyrex at low temperatures; graphite and Al_2O_3 at high temperatures; avoid siliceous material at high temperatures	Under charcoal or halide fluxes Under potassium chloride or charcoal; argon or vacuum
	In	Porcelain, pyrex, mullite	Hydrogen flame or argon
	Tl	Porcelain, pyrex	Hydrogen
Group IV	Si	Silica, TiO_2, ZrO_2, ThO_2	In argon, vacuum, or helium
	Ge	Graphite, silica, Al_2O_3	Nitrogen or vacuum
	Sn Pb	Pyrex, porcelain, graphite, spinel, mullite	Under charcoal; hydrogen
Pnictogens	As	Pyrex, silica	Low as under halide flux, high as in sealed tubes
	Sb	Porcelain, graphite, silica	Under charcoal; hydrogen
	Bi	Pyrex, porcelain, silica, graphite, CeS	Hydrogen
Chalcogens	Se Te	Silica or graphite	In argon or vacuum
Sc, Y, and rare-earth metals	Ce Sc Y La Pr, etc.	CeS Graphite jacketed tantalum or CaO or BeO	Argon or vacuum

Bibliography

[1] https://science.osti.gov/bes/Community-Resources/Reports/
[2] P. Ball, *MRS Bull.*, **42**, 698 (2017).
[3] C. Herring, *Phys. Rev.*, **52**, 365 (1937).
[4] W. Pauli, *Letter from Pauli to Rudolf Peirels* (Sept. 29, 1931).
[5] R. J. Lewis (ed.), *Hawley's Condensed Chemical Dictionary* (Wiley, 2016).
[6] *The Mystery of Matter: Search for the Elements*, DVD film released for Public Broadcasting System (PBS) (2015).
[7] W. Hume-Rothery, *The Metallic State, Electrical Properties and Theories* (Oxford University Press, 1931).
[8] J. F. Nye, *Physical Properties of Crystals: Their Representation by Tensors and Matrices* (Oxford University Press, 1985).
[9] B. R. Pamplin (ed.), *Crystal Growth* (Pergamon Press, 1980).
[10] J. H. Westbrook, *Intermetallic Compounds* (Wiley, 1967).
[11] H. B. Cao *et al.*, *Phys. Rev. B.*, **93**, 134423 (2016).
[12] A. Bannerjee *et al.*, *Science*, **256**, 1055 (2017).
[13] M. A. McGuire, H. Dixit, V. R. Cooper, and B. C. Sales, *Chem. Mat.*, **27**, 612 (2015).
[14] B. Huang *et al.*, *Nature*, **546**, 270 (2017).
[15] T. Song *et al.*, *Science*, **360**, 1214 (2018).
[16] P. C. Canfield, and Z. Fisk, *Phil. Mag. B.*, **65**, 1117 (1992).
[17] M. G. Kanatzidis, R. Pottgen, and W. Jeitschko, *Angew. Chem.*, **44**, 6996 (2005).
[18] D. E. Bugaris, and H. C. zur Loye, *Angew. Chem.*, **51**, 3780 (2012).
[19] M. Isobe, and Y. Ueda, *J. Phys. Soc. Jpn.*, **65**, 1178 (1996).
[20] M. D. Lumsden, B. C. Sales, D. Mandrus, S. E. Nagler, and J. R. Thompson, *Phys. Rev. Lett.*, **86**, 159 (2001).
[21] B. C. Sales, M. D. Lumsden, S. E. Nagler, D. Mandrus, and R. Jin, *Phys. Rev. Lett.*, **88**, 095901 (2002).
[22] F. Heusler, *Z. Angew. Chem.*, **17**, 260 (1904).
[23] E. Adams, W. M. Hubbard, and A. M. Syelles, *J. Appl. Phys.*, **23**, 1207 (1952).
[24] M. A. McGuire, H. Cao, B. C. Chakoumakos, and B. C. Sales, *Phys. Rev. B.*, **90**, 174425 (2014).
[25] T. J. Williams *et al.*, *Appl. Phys. Lett.*, **108**, 192403 (2016).
[26] N. H. Dung *et al.*, *Adv. Mat.*, **1**, 1215 (2011).
[27] H. Fujii, T. Hokabe, T. Kamigaichi, and T. Okamoto, *J. Phys. Soc. Jpn.*, **43**, 41 (1976).
[28] B. C. Sales, M. A. Susner, B. S. Conner, J.-Q. Yan, and A. F. May, *Phys. Rev. B.*, **92**, 104429 (2015).

[29] M. B. Stone, M. D. Lumsden, S. E. Nagler, D. J. Singh, J. He, B. C. Sales, and D. Mandrus, *Phys. Rev. Lett.*, **108**, 167202 (2012).

[30] B. C. Sales, A. F. May, M. A. McGuire, M. B. Stone, D. J. Singh, and D. Mandrus, *Phys. Rev. B.*, **86**, 235136 (2012).

[31] J. Ma *et al.*, *Nat. Nano*, **8**, 445 (2013).

[32] B. C. Sales, A. S. Sefat, M. A. McGuire, R. Y. Jin, D. Mandrus, and Y. Mozharivskyj, *Phys. Rev. B.*, **79**, 094521 (2009).

[33] H. B. Cao *et al.*, *APL Mat.*, **3**, 062512 (2015).

[34] O. Delaire *et al.*, *Phys. Rev. B.*, **91**, 094307 (2015).

[35] J. Q. Yan, M. A. McGuire, A. F. May, H. Cao, A. D. Christianson, D. G. Mandrus, and B. C. Sales, *Phys. Rev. B.*, **87**, 104515 (2013).

[36] J. Q. Yan, B. C. Sales, M. A. Susner, and M. A. McGuire, *Phys. Rev. Mat.*, **1**, 023402 (2017).

[37] M. A. McGuire *et al.*, *Phys. Rev. Mat.*, **1**, 064001 (2017).

[38] B. C. Sales *et al.*, *NPJ Quantum Mat.*, **2**, 33 (2017).

[39] M. A. Susner *et al.*, *ACS Nano*, **9**, 12365 (2015).

Chapter 2

Chemical Bonding and Structural Relationships in Extended Solids

Paul H. Tobash[*,†] and Svilen Bobev[*,‡]

*Department of Chemistry and Biochemistry,
University of Delaware, Newark DE 19716, USA
†Los Alamos National Laboratory, MST-16:
Nuclear Materials Science
Los Alamos, NM 87545, USA
‡bobev@udel.edu

2.1. Introduction

The realm of solid-state chemistry and condensed matter physics offers the scientist opportunities for discovering vast amounts of novel compounds that exhibit a myriad of structural entities. New materials continue to be made, often times serendipitously, and illustrate that the growing number of phases and polytypes in already busy-phase diagram landscapes, frequently challenge our chemical intuition. These difficulties not only include finding members of a particular system which were thought to be nonexistent based upon formation relating to periodic table trends using simple ball and stick packing models, but also what governs the large number of ways in which the atoms can be arranged or "adapt" to their surroundings in the crystal structure. Indeed, these mentioned points both remain open questions for intermetallic compounds which are manifested through the following: (1) moving away from serendipitous material discovery and total control of reaction outcome; (2) being able to rationalize the formation of one structure over another; and (3) the characteristics of the structure which lead to the observed physical properties. Of course the way in which the atoms in a crystal structure are arranged will have a significant effect on the observed physical properties, many of which are technologically relevant such as superconductivity, magnetic refrigeration, and spintronics, so comprehending the structure–property relationships will serve to improve and ultimately design better materials to utilize in our ever-changing world.

Quantum materials become increasingly important in all aspects of our daily lives. This rather broad class of materials has its roots in everything we rely on in lives and occupations that include everything from electronics and computational technologies to energy usage and storage [1, 2]. These materials can generally be defined as being shown to exhibit exotic properties where electron correlations are at the forefront for these properties to be manifested [3]. The fundamental understanding for how the phenomena arises stems from the "collective" nature of the interactions of the electrons in the system. These electron states have masses that are much greater than the singular electron in the system. Sometimes this is referred to as being emergent behavior and becomes very relevant when exploring systems where one may have partially filled d- or f-electronic shells, structures with reduced dimensionality, and/or narrow bandwidths. As we will see later on, the consequence of these correlations is displayed in a variety of forms that include, but is not limited to heavy-fermion behavior, superconductivity, complex magnetism, Kondo physics, topological materials, and spin liquids. The recognition of these unusual states in quantum materials can be easily assessed from a number of low-temperature thermophysical property measurements that may include, e.g., magnetic susceptibility, specific heat, and electrical resistivity. Many times in condensed matter research it is said that "new materials" will result in "new physics" and this will ultimately drive the research in a macroscopic way that extends beyond to the broader scientific community. Taking the phenomena of unconventional superconductivity, for example, even though the critical temperatures required for superconducting transitions reach nearly halfway to room temperature, we still do not yet have a firm grasp on the origins of unconventional superconductivity in intermetallic and oxide compounds. Examples of superconductivity in compounds are numerous and for the majority of cases, its emergence is serendipitous, thereby hampering the possibility of tunable control over physical properties. Despite this elusiveness, common structural trend can often be extracted from several families of compounds and provides the materials scientist with a way to rationalize structure–property relationships. It turns out when this mindset is coupled with theoretical endeavors, this approach becomes very powerful especially with the supercomputers available to carry out these long and often exhaustive electronic structure calculations [4–6]. Take for instance the past 3–4 decades of research where waves of similar materials were explored, such as the high-temperature oxide superconductors in the 1980s with CuO planes as a common building block in a large family of interrelated compounds, $M\mathrm{In}_3$ (M = Ce, Pu) blocks in the popular "115"-type compounds (many of them showing superconducting transitions and correlated electron behavior), and more recently the iron-based superconductors (FeSCs), boasting Fe-containing slabs of common topologies, exhibiting a close relationship and competition between pseudogap phases, spin density waves, and nematic phases in the materials.

Another peculiar case worthy of mention here is the compound MgB_2 that was structurally identified more than six decades ago [7], but only recently found to have a superconducting transition at 39 K [8]. After its discovery and stir in the scientific community, many began to wonder what other combinations of elements, of course being structurally related to MgB_2, would display the same behavior. Oddly enough, the structure of this binary compound is very simple, containing hexagonal sheets of three-bonded boron atoms assembled by the fusion of $\frac{1}{\infty}[\mathrm{B}_2]$ zig-zag chains resembling a "honeycomb-like" motif exactly like in graphite. The Mg atoms are intercalated between these planar boron sheets in the structure. Importantly, the layered structure has implications for the observed superconductivity in which the two groups of π and σ electrons form distinct pairing states which coexist in the material [9]. Recognizing this, it comes as no surprise that $\mathrm{YbGa}_{1.1}\mathrm{Si}_{0.9}$ was found to superconduct [10], after all, the structure contains $\frac{1}{\infty}[(Ga_{1-x}Si_x)_2]$ layers identical in topology to those in MgB_2. Although this layered structural motif is very common for many binary and ternary intermetallic compounds [11], the reason why only a few exhibit the same superconducting behavior remains a question to be answered.

In just the recent couple of years, the exciting world of a special field known as skyrmionics (offering a large impact on future processing devices and magnetic storage) have entered into the picture and offer yet another novel electronic state that can be presented in materials that lack inversion symmetry [12–14]. They can be experimentally observed in a number of characterization techniques such as neutron diffraction, Lorentz transmission electron microscopy, and scanning tunneling microscopy. Some examples include the topological Hall effect being observed in materials such as MnSi [15, 16], FeGe [17], and MnGe [18].

The discussion that follows will look more specifically at examples where structure and bonding play an important role in the emergence of various thermophysical behavior. We will begin the discussion with the following topical areas: (1) the influences that polymorphic compounds (materials with the same formula but crystallized with distinct structures) and atomic size effects have on the physical properties; (2) crystallographic disorder in compounds and its effect on properties; (3) aspects of structure and bonding in materials; (4) an overview of the some structural trends in germanides, which have been studied in our laboratory for over a decade; and (5) some examples of quantum materials exhibiting heavy-fermion behavior, including some exotic Pu-containing materials with correlated $5f$-electronic states.

2.2. Structure and Chemical Composition

Polymorphism within intermetallic compounds is a recurring theme in which the list of new polymorphic forms continues to grow. These are compounds that are composed of the same elements and crystallize with identical formulas but with different structures (much like isomerism in the molecular state). In these cases, depending on the type of experimental conditions used (either the flux growth technique or the classic solid-state method) can bring about the isolation of specific polymorphs leading in one case to the more thermodynamically stable product while another brings about the formation of a metastable or kinetic product. These are commonly identified with a designated nomenclature, in which the a-form is usually given to the low-temperature product in the reaction while the b-form is given to the high-temperature product [19]. In fact, the occurrences of polymorphism in the literature abound, like β-$RENiGe_2(RE = RE =$ Rare-Earth metal) [20] and β-SiB_3 [21], which have been isolated from flux growth reactions in liquid indium and gallium, respectively. Other cases include the new forms of $YbAlB_4$ [22, 23] and $CeNiSb_3$ [24] as well as $CaGe_2$ [25] and RE_3Ge_5 $(RE =$ Sm, Tb, Dy) [26, 27], all representing new forms to those known previously. In addition, two other binary germanides, La_5Ge_3 and Ce_5Ge_3, have been found in tetragonal forms [28], adding new knowledge to the abundant family of "5–3 phases" which most commonly crystallize with the hexagonal Mn_5Si_3-type [11]. The latter materials can also exemplify the way in which electronic correlations can be understand from the paramagnetic behavior of Ce^{3+} electronic ground state in Ce_5Ge_3 with no long-range magnetic ordering down to 5 K and Pauli-paramagnetism exhibited in La_5Ge_3. In this case, the La-analog may be used to subtract phonon contributions to the specific heat and ultimately extract the magnetic effects in Ce_5Ge_3.

Speaking of binary phases, the published-phase diagrams are a good starting point for becoming familiar with the phases formed between two or three respective elements in a given system [19]. The information contained within them has been compiled through extensive studies on the conditions such as composition, temperature, or pressure, needed to synthesize a specific compound. However, several instances of novel compounds which were previously unknown to their respective systems, or known compounds with certain aspects of their structure which have been overlooked, offer the possibility of a reexamination for their chemistry and ultimately a clearer understanding for their formation. The former conundrum is manifested by either the completely new phases prepared from the respective elements, or the identification of metastable compounds (often polytypes/polymorphs of known phases), which can only be accessed by employing different synthetic methods. A good example here is the already mentioned $RE_3Ge_5 (RE = Sm, Tb, Dy)$ [26, 27], which is a structurally reassessed variant of some of the myriad of compounds described in the literature with the general formula $REGe_{2-x}$ [11]. We will note explicitly that this is more of a rule, rather than exception in these systems, since the majority of "1–2 phases" are not line compounds [29–31]. In other words, materials with the formula $REGe_2$ rarely exist, instead, compositions where the elemental ratio is noninteger abound (and often time show characteristics of structure and bonding which appear to be incorrect). Careful synthesis and structural characterization, later discussed in the structure and bonding section, reveals that the variations "x" in the formula $REGe_{2-x}$ can be controlled, and as it will be shown, this will be inherently related to small subtleties in the structure and bonding, ultimately leading to a reevaluation of structure–property correlations in opposition to what previous studies have originally reported. For example, Sm_3Ge_5 (= $SmGe_{1.67}$, i.e., $SmGe_{2-x}$ for $x = 1/6$) exhibits two polymorphic forms — one crystallizing with the orthorhombic Y_3Ge_5-type (space group $Fdd2$) while the other one crystallizing with a novel hexagonal type (space group $P\bar{6}2c$). Each one shows a distinct long-range antiferromagnetic ordering temperature that corresponds to $T_N = 30$ K for α-Sm_3Ge_5 and $T_N = 10$ K for β-Sm_3Ge_5. It turns out that the synthesis of each compound may be controlled depending on the reaction cooling rates. This occurrence is related to the more thermodynamically stable product and the metastable phases.

More recently, a subsequent work on $REGe_{2-x}$ (for $x = 1/4$, $RE =$ La-Nd, Sm) shows a different type of vacancy-ordered variants with the formula RE_4Ge_7 [32]. These materials, obtained via crystal growth conditions under flux, also crystallize in what can be considered a "superstructure" of well-known type (α-$ThSi_2$-type [11]), where the crystallographic long-range and/or short-range ordering was unveiled using transmission electron microscopy. Ultimately, when a comparison of the series is done between the later rare earth atoms' structure and bonding and the variants of the α-ThSi$_2$-type alloys, an assessment of the intrinsic effects, the vacancies, and the rare-earth atom has the thermodynamic properties.

The importance of synthesis control is also on display for the line compounds, for instance, the "1–1–4" materials REAlB$_4$ (RE = Yb, Lu) [22, 23]. Macaluso *et al.* showed that using an excess of Al allows for the formation of two polymorphic forms, which are easily recognizable from the crystallographic habit between the plate-like and needle-like morphologies that corresponded to β-YbAlB$_4$ and α-YbAlB$_4$, respectively. It turns out that the Yb/Al layers in both structures are very similar and both are made up of heptagonal rings where the Yb atom is centered and pentagonal rings where the Al atom is centered. The polymorphism only occurs for the way in which these layers are packed in the structure. This of course also has a dramatic effect on the exhibited physical properties where a largely different Sommerfeld coefficient is realized thereby classifying them as heavy-fermion materials with the β-phase having a value of $300\,\text{mJ/mol-K}^2$ and the α-phase with a value of $130\,\text{mJ/mol-K}^2$.

Another rich example of polymorphism in intermetallics can be pointed to the binary actinide-containing PuGa$_3$ compounds (rhombohedral and hexagonal crystallographic forms) [33]. Although the "1–3" stoichiometry is preserved in PuGa$_3$, it is not isostructural to its neighbor Group 13 cubic PuIn$_3$ and Group 14 cubic PuSn$_3$ compounds. Instead, PuGa$_3$ can be isolated in two polymorphic forms depending on the synthetic conditions that have been described nicely by Boulet and coworkers. The rhombohedral (high temperature form) crystallizes with the space group $R\bar{3}m$ (No. 166) and a hexagonal variant (low temperature form) crystallizes with the Ni$_3$Sn structure type with the $P6_3/mmc$ space group (No. 194) [11]. The former modification of PuGa$_3$ can be made in polycrystalline form from annealing at high temperatures. Of course, this difference in structures has important implications for the low temperature behavior displayed by both compounds. Both the heat capacity and susceptibility data show that the hexagonal form of PuGa$_3$ undergoes an antiferromagnetic transition at $T_N = 25\,\text{K}$. When the low-temperature C_p/T heat capacity data is plotted, and a linear fit is done below $10\,$K, where it is already in the magnetic state, this analysis gives a Sommerfeld coefficient of $199\,$mJ/mol-K^2 with $\beta = 1.26\,$mJ/mol-K^4. This value then gives a Debye temperature of $221\,$K. Strong electronic correlations are also suggested to be present in the rhombohedral form of PuGa$_3$ with a Sommerfeld value of $100\,$mJ/mol-K^2 in the ferromagnetic state. When the magnetic susceptibility data for the latter is fitted to a modified Curie–Weiss law $\chi(T) = C/(T - \theta_{CW}) + \chi_0$, this gives an effective moment of $0.80\ \mu_B$ for the Pu^{3+} ($0.84\ \mu_B$ expected) with $\theta_{CW} = 14\,$K and $\chi_0 = 4.4 \times 10^{-4}\,$emu/mol.

The discovery of new compounds can also be brought about from the substitution of "chemically similar" elements, i.e., elements belonging to the same periodic group, through the replacement of one element in the crystal structure for another while maintaining an isostructural system. Notably, this will also be later shown to have significant implications on the observed physical properties, which is demonstrated for a large number of phase relationships in $RE_5Tt_4(RE = \text{La–Er}, Tt = \text{Si, Ge})$ compounds for example [34–56]. Once recognized, the formation trends across a particular system will allow for the establishment of the end-members that define the series of families. A comparison of the structure and bonding patterns across the isostructural members will not only provide an assessment of the crystal structure in the way it is assembled but more importantly what structural effects accompany the substitution, affording a rationalization as to why the atoms occupy specific sites in a structure. Take for example Gd$_5$Si$_2$Ge$_2$, a ternary form of RE_5Tt_4 ($RE = $ Gd, $Tt = $ Si, Ge) that is "chemically tuned" and exhibits a giant magnetocaloric effect [57]. This is a common approach in intermetallic compounds, where chemically similar elements are used to manipulate the electronic states of the materials, is employed to influence the observed properties. In this particular case, both Gd$_5$Si$_4$ and Gd$_5$Ge$_4$ are known, however, it is only their intergrowth that shows such unique properties. One can envision the latter as being built from distinct slabs, which account for crystallographic transformations closely associated with its properties. These include the magnetic coupling of the $4f$ moments in the layers that influence the type of magnetic ordering that is observed, chemical doping on the lanthanide site can alter not only the crystal structure, but also the long-range magnetic order, and lastly the electronic correlations are more largely affected in the structure rather than the size effects. The outcome of rational substitutions in the parent structure can led to a number of quantum materials that exhibit for example colossal magnetoresistance [58], giant magnetostriction [35, 46], spontaneous voltage generation [43], and giant magnetocaloric behavior [57], depending on the crystallographic site that is targeted. All of the above underscores the need to find and understand the close relationship between the structural parameters that give rise to the electronic correlations, and how they are coupled together, i.e., identifying what "structural knob" needs to be turned in the crystal structure in order to modify the desired properties.

Speaking of chemical substitutions/tunability, it is important to discuss atomic site preferences, or as it is sometimes termed, "coloring" of the atomic sites in extended solids. Besides the number of "experimental endeavors" with regard to the already mentioned RE_5Tt_4 (RE = La–Er, Tt = Si, Ge) systems, many other studies have been done and some theoretical approaches have been developed. Such work on model systems of $LaNiGe_2$ [59], $CaAl_2Si_2$ [60], $CaBe_2Ge_2$ [61], $BaAl_4$ [62] and $ThCr_2Si_2$ [63], among others, has been particularly successful. Several of these studies have noted important factors like electronegativity, atomic sizes, or coordination numbers of the constituent atoms, which govern for a large part where they will be located in the structure. Ultimately, the information gathered from studies like these will permit for the necessary connection between the basic concepts of bonding and composition. These ideas have been discussed in great detail in a review by Miller, in which he describes that the two factors that dominate the structural landscape are the site energy and the bond energy [64]. The first refers to the valence electrons' energetic contribution from the constituent atoms which is essentially the net charge at each crystallographic site, while the second takes into account the optimization of the bonding.

When "chemically different" elements are being studied (that is to say elements which do not belong to the same group in the periodic table), the ability to rationalize site preferences becomes a bit easier. However, sometimes structure-directing factors which are suggested to play a large role for the formation of a structure turn out to be not as clear cut. This is embodied when elements from the same group coexist in the crystal structure and because the electronic requirements will be the same for both, the preference for specific atomic arrangements will be defined by other reasons. Indeed, there are very few cases where a site preference for two elements in the same group is manifested. These include compounds like Ba_2BiSb_2 [65] having fused six- and four-membered rings where the Bi atoms reside on four-bonded positions and Sb atoms on three-bonded positions, $K_2Bi_{8-x}Sb_xSe_{13}$ [66] containing Bi atoms and Sb atoms on specific sites, $GdCuAs_{1+x}P_{1-x}$ [67] with As atoms and P atoms located on square or puckered sheets, respectively. Work on compounds containing Ge and Pb, such as $(Sr_xEu_{1-x})_2Ge_2Pb$ [68], and the Ge and Sn, such as $RESn_{1+x}Ge_{1-x}$ $(RE = Gd-Tm, Y)$ [69], $RE_2[Sn_xGe_{1-x}]_5$ $(RE = Nd, Sm)$ [70] and $RE[Sn_xGe_{1-x}]_2$ $(RE = Gd, Tb)$ [70] also provides evidence for preferential distribution among the group-14 elements. In the $RESn_{1+x}Ge_{1-x}$ structure for example, the Ge atoms are situated on $\frac{1}{\infty}[Ge_2]$ zig-zag chains and Sn composes two-dimensional $\frac{2}{\infty}[Sn]$ square sheets. Notably, these site preferences are easily recognized from the refinements of the crystal structure using conventional single-crystal X-ray diffraction methods, but the importance of the availability of quality crystalline material to carry out such studies cannot be overstated. The crystallographic sites will either be occupied separately by the two chemically similar atoms or in some cases the atomic site can be refined as a mixture of both elements but shows a much larger occupation factor of one over the other. These cases are markedly different than crystal structures which display a more random or statistical distribution of atoms is observed as that demonstrated by $Nb_5(Ge_xSn_{1-x})_2$ [71], $Ba_6(Sn_xGe_{1-x})_{25}$ [72], and $Ca(Sn_xGe_{1-x})_2$ [73].

2.3. Structure, Chemical Bonding, and Structure–Property Relationships

Even a cursory view of the compounds formed by the rare-earth or alkaline-earth metals of the periodic table along with various p-block elements (those of groups 13–16) only demonstrates the myriad of structural arrangements which can be realized. These can range from the formation of metal clusters [74–81] (usually resulting from electron deficient systems) to extended sheets or layers [82–88] (usually connected with electron-rich bonding arrangements) to three-dimensional networks [89–99]. The type of arrangement which is realized is mostly dependent on reasons related to electronic effects or atomic size limitations which results in the most efficient crystal packing in the structure. Very often the crystal chemistry of the structure can be understood by carefully recognizing the simple building motifs for the way in which a structure can be assembled. This can be a common structural thread that can be recognized in related compounds no matter how complex the structure may seem. For example, building moieties may be connected together, stacked on top of one another, or have additional fragments inserted or inter-calated within the structure of the existing framework. All of these outline some of the approaches employed for the synthesis of solid-state structures have been discussed at great length by DiSalvo [100] and Jansen [101]. Importantly, the majority of the synthetic rationale that these authors put forward involves the discovery of several more structural topologies and bonding arrangements. One fundamental approach is being able to reduce some structures down to a "basic topology" such as building blocks being described as squares, nets, tetrahedral fragments. In some cases, a resulting homologous series can be described and can be ultimately used to generate or "foresee" several other compounds yet to be discovered. As an example, the structures of Ce_2Sn_5 [102] and Ce_3Sn_7 [103], described with the formula RE_nSn_{3n-2}, can be easily constructed from the stacking of $CeSn_3$ units (easily recognizable as the ubiquitous perovskites with general formulas ABO_3 boast the same structural fragment). Also of noteworthy importance is the large family of ternary germanides discovered herein which include $RESn_{1+x}Ge_{1-x}$, (RE = Gd–Tm, Y) [69], $RE_2[Sn_xGe_{1-x}]_5$ (RE = Nd, Sm) [70], and $RE[Sn_xGe_{1-x}]_2$ (RE = Gd, Tb) [70] and can be assembled from zig-zag or square sheet fragment moieties. Another suitable for referencing here (and growing) family of inter-related compounds are the Pu-containing materials PuX_3 [104], $PuCoX_5$ [105], Pu_2PtX_8 [104], $PuPt_2X_7$ [106] (X = Ga or In). When the individual building blocks are examined independently, other possible stoichiometries can be hypothesized and theoretical calculations must be carried out in order to predict if the compound would be thermodynamically stable.

With the advancement of our computational tools as well as the recent push in applying machine learning principles for the prediction of new intermetallic compounds, the phase space that can be explored can be done a lot more quickly and efficiently [106–108]. This also requires the computations to have defined parameters and boundary conditions in order for this optimization to work well, as detailed by Jansen. Besides the prediction of controlled structural topologies being difficult, there are additional challenges to structure and bonding when including factors such as defects, vacancies, or grain boundaries. Microstructure and defects induced from radiation damage, such as those often found in Pu intermetallics, can too have a dramatic effect on properties [109, 110].

When comparing the structural chemistry across families of compounds of a given system it is often times useful to recognize bonding arrangements which are identical or that slightly resemble one another either through a shifting of the building fragments or substitution of a "chemically similar" element. Particularly for the germanides, the formation of one dimensional Ge zig-zag chains described as $\frac{1}{\infty}[Ge_\infty]$ is a recurring theme both in binary and ternary phases. In fact, this topology is so common that it is observed for all of the monogermanides crystallizing with the formula $REGe$ (RE = La–Nd, Sm–Y) [11]. Furthermore, depending on the way in which these chains are arranged can lead to two-dimensional layers or three-dimensional networks as depicted below in Scheme 2.1. These two arrangements which will be referred to as "parent" structures are commonly seen for rare-earth binary and ternary compounds such as

REGe

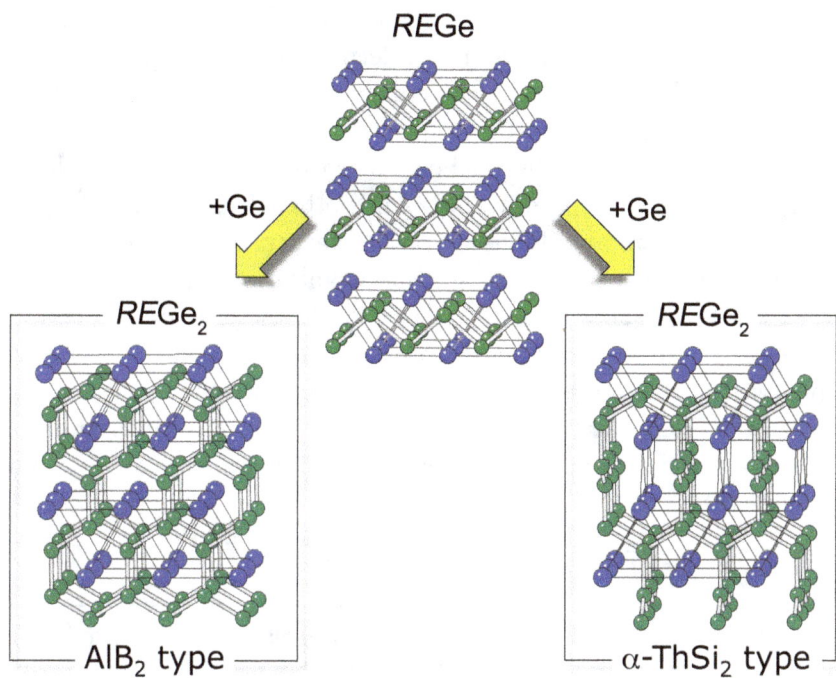

+Ge

REGe$_2$

+Ge

REGe$_2$

AlB$_2$ type

α-ThSi$_2$ type

Scheme 2.1. Schematic representation of how the structure of the rare-earth metal monogermanides *RE*Ge (top) can be related to the structures of two of the most common types among the rare-earth metal digermanides *RE*Ge$_2$. Stacking slabs of Ge atoms forming zig-zag chains in parallel fashion leads to the formation of two-dimensional planar sheets in *RE*Ge$_2$ (known as the hexagonal AlB$_2$ structure type), while a formation of a three-dimensional network results from the perpendicular fusion of germanium zig-zag chains (an arrangement that is known as the tetragonal α-ThSi$_2$ structure type).

the ones discussed specifically within the chapters herein along with a number of others mentioned briefly in the following paragraphs.

As an example, the two arrangements will be explained in better detail below. As depicted in Scheme 2.1, the fusion of parallel zig-zag chains results in a two-dimensional layer reminiscent of a honeycomb-like topology, i.e., referred to as AlB$_2$-type. Within these planar nets, the Ge atoms are three-bonded to one another and are identical to the coordination environment of the carbon atoms in the well-known graphite structure. In the case where these chains are not parallel but instead are fused along perpendicular directions, this arrangement gives rise to again three-bonded Ge atoms which create a three-dimensional network, i.e., referred to as the α-ThSi$_2$-type. The rare-earth cations are then either intercalated between these layers or in the channels created by the network, respectively.

It is now suitable to discuss how these two structures are realized across the lanthanide family. This will be beneficial since all(most) of the binary compounds discussed herein can be derived in some manner from their "parent" types. Importantly, the type of structural arrangement which is observed is largely dependent on the ratio of the ionic size of the rare-earth cation and the Group 14 element along with the number of defects, i.e., vacancies on the tetrel atom sites in the crystal structure. As shown in Scheme 2.2, the prominent structure types are outlined for the binary silicides, germanides, and stannides [137].

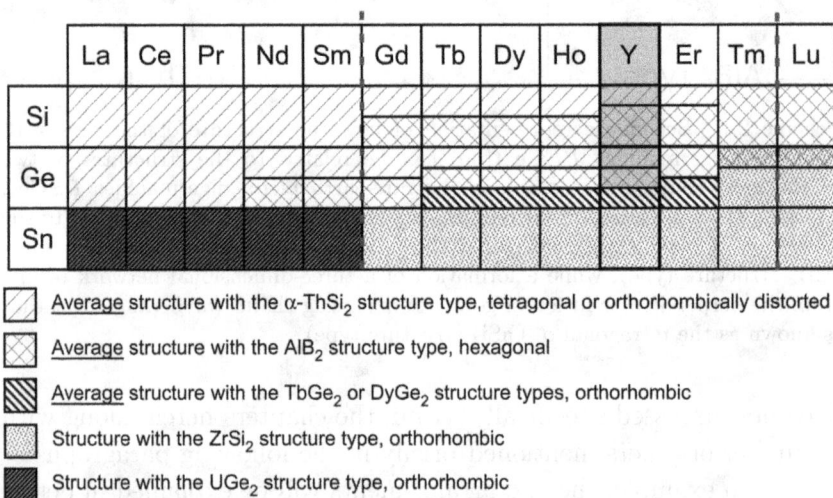

	La	Ce	Pr	Nd	Sm	Gd	Tb	Dy	Ho	Y	Er	Tm	Lu
Si													
Ge													
Sn													

▨ Average structure with the α-ThSi$_2$ structure type, tetragonal or orthorhombically distorted

▨ Average structure with the AlB$_2$ structure type, hexagonal

▨ Average structure with the TbGe$_2$ or DyGe$_2$ structure types, orthorhombic

▨ Structure with the ZrSi$_2$ structure type, orthorhombic

▨ Structure with the UGe$_2$ structure type, orthorhombic

Scheme 2.2. Overview of the structural trends among binary silicides, germanides, and stannides with chemical formulae RESi$_{2-x}$, REGe$_{2-x}$, and RESn$_{2-x}$, respectively, where $x = 0$ or is varied but typically $x < 1/3$. The red-dashed vertical lines represent the missing Eu and Yb, which are nominally divalent in these cases and do not follow the established periodicity for the remaining, nominally trivalent member of the lanthanide series. Yttrium is included in the sequence and its arrangement is based on its atomic size.

Particularly with the focus of this study on the germanide compounds, the focus will be on the α-ThSi$_2$ and AlB$_2$-types which represent the majority of "parent" structures that the REGe$_{2-x}$ phases can be derived from. From the trends, the type of rare-earth cation largely has an influence on the structure and stoichiometry that is formed. For example, the largest rare-earth cations in the series (lighter in mass) usually form with the α-ThSi$_2$-type structure [11]. Once the cationic size becomes gradually smaller due to the lanthanide contraction, the formation of AlB$_2$-type compounds for the heavy lanthanides is observed [11].

The nonstoichiometry in this class of compounds is due to the creation of Ge vacant sites in the crystal structure. The defects formed, pose a problem for understanding of structural and bonding patterns which in turn hampers the derivation of proper structure–property relationships. These defects (or vacancies), which can be ordered throughout the crystal lattice, lead to the existence of complicated structures that can be described using supercell space groups. Due to the large breadth of stoichiometries that can form, reassessment of the rare-earth germanides will be beneficial. A simple schematic can be referred to below which outlines the manner in which the creation of long-range ordering of vacant sites in the crystal structure leads to the formation of a supercell to describe the new structure. As can be seen in Scheme 2.3, the unit cell of a lattice having no vacant sites can be inscribed and represents the smallest repeating unit of the lattice. If, however, vacant sites are created (represented by the open circles in Scheme 2.3) and are in addition ordered in some way, i.e., in this case the vacant site occurs every other atom, a new unit cell must describe the structure to account for the change in the lattice. As can be seen, this new unit cell is four times larger ($\sqrt{2}a \times 2\sqrt{2}a$) compared to the original and now represents the smallest repeating unit of the structure. Examples of this type of chemistry can be found with a number of binary germanides that crystallize with the formula RE_4Ge$_7$ [32]. A number of other reports have also noted the long-range ordering of vacancies in rare-earth germanides [112–115].

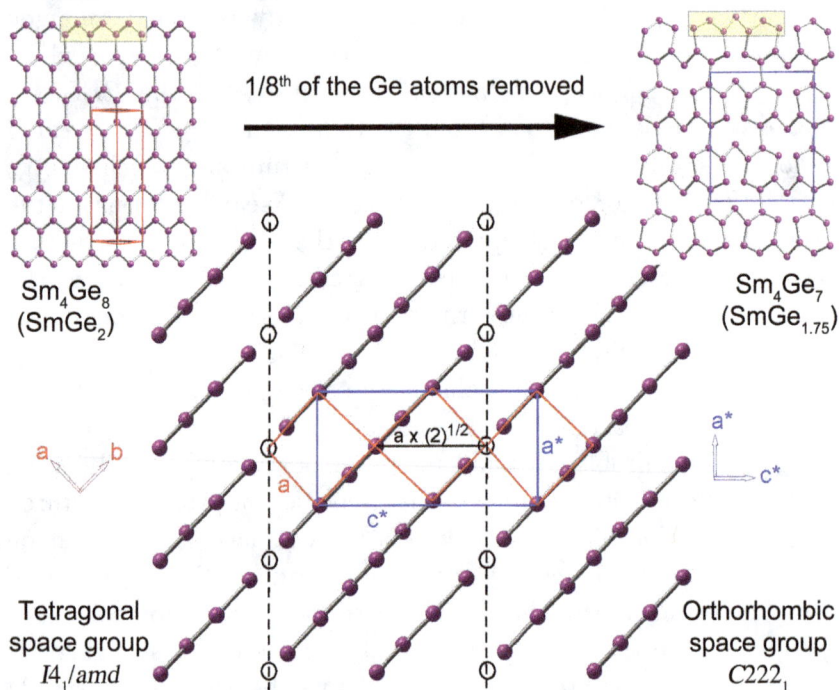

Scheme 2.3. General schematic demonstrating a subcell–supercell relationship. The choices of the respective unit cells are designated in red and blue. The specific example is based on the reportedly nonstoichiometric $SmGe_{2-x}$ ($x \approx 1/8$) with an average structure adopting the α-$ThSi_2$-structure type. Controlling the chemistry and the number of vacancies allows for the long-range order of the latter and the formation of a new structure type via the described distortions and symmetry transformations.

We will end our discussion in this section focusing on the structures of several compounds, noting some examples illustrating the abundant structural chemistry of the group-14 element Ge. This is because all of the chemistry described herein can be drawn upon and related to numerous compounds that exhibit similar structural arrangements already in the literature. Importantly, we want to show the diversity of this chemistry by categorizing in groups representing one, two, and three dimensionalities of the Ge subnetwork. The simplest and most fitting way to begin would be to point out structures that contain isolated Ge atoms within the crystal structure. A great example is in $La_{11}In_6Ge_4$ containing discrete Ge atoms located in the 16-membered and 4-membered tunnels of the In network [116]. The Ge–Ge distances are very long, measuring *ca.* 3.77 to 3.985 Å and essentially can be formulated to have a formal oxidation state of negative 4 (i.e., a fully reduced Ge atom with no covalent bonds to its neighbors needs four electrons to complete its octet). This compound is a close relative to the $Ho_{11}Ge_{10}$ structure [11], in that it can be derived from the substitution of the Ge_4 squares and Ge_2 dimers with In, resulting in the formation of the three-dimensional In network [117]. Normal Ge–Ge distances indicating bonding interactions measure in the range of 2.5–2.8 Å, which is slightly longer than that observed for elemental Ge and the sum of their covalent radii according to Pauling [118]. Also included here is the compound La_3In_4Ge containing a layered network of In atoms which are sandwiched between layers of La and isolated Ge atoms [119]. Other examples are La_3InGe and Ca_5Ge_3, the latter of which contains lone Ge^{4-} atoms along with isolated dimers making Ge_2^{6-} units [120].

When two Ge atoms come in close contact with one another, the formation of dimer units Ge_2^{6-} (isoelectronic to the Br_2 molecule) is seen. Some examples include the Ge–Ge dimers in Ca_5Ge_3 [120] or in the planar two-dimensional layers in RE_2InGe_2 and RE_2MgGe_2 (RE = Sm, Gd–Ho, Yb) [121, 122], Yb_4MgGe_4 [123] or $RE_{5-x}Mg_xGe_4$ (RE = Gd–Tm, Lu, Y) [124] of which the latter two contain dimerized units, fused together with either four coordinate In or Mg atoms resulting in $[InGe]_2$ or $[MgGe]_2$ sheets. These Ge–Ge bonds measure *ca.* 2.5–2.6 Å, in good agreement with a number of other rare-earth binary and ternary germanide compounds [125–128]. Of special note are the two kinds of Ge dimers which are observed in Yb_4MgGe_4 and $RE_{5-x}Mg_xGe_4$ compounds [123, 124], one of which was just mentioned while the other dimers are found residing between the $[MgGe]_2$ layers. In fact, a number of compounds which crystallize with this structure have been well characterized and studied due to the bond making/breaking processes which can be induced with these interslab Ge–Ge dimers leading to unique physical properties [34–56]. This has been accomplished when the structure is subjected to atom substitution, temperature, pressure, and magnetic field that induces structure transformations.

The two-dimensional layers formed by the fusion of two zig-zag chains, the same sort of topology as displayed in Scheme 1 containing three-bonded Ge atoms can be recognized in a number of binary compounds. Some noteworthy examples include the hexagonal sheets in RE_3Ge_5 (RE = Sm, Gd−Dy) [26, 27], Y_3Ge_5 [129], or the planar nets in Yb_3Ge_5 [130], all of them containing honeycomb-like Ge layers. Still we can see the formation of another topology giving rise to the formation of two-dimensional layers which form a square lattice. This again can be envisioned as being built from the fusion of several zig-zag chains which come in close contact. The Ge atoms, $^2_\infty[Ge]$, in these fragments are four-bonded and are only realized for compounds containing smaller metal cations such as that of $ScGe_2$, $ZrGe_2$, $TmGe_2$, $LuGe_2$, $ThGe_2$, and UGe_2 [11]. Some other unusual examples include compounds such as $TmGe_{1.9}$ [131], Er_2Ge_5 [132], or the $RESn_{1+x}Ge_{1-x}$ [69] and related compounds explained in detail later on in the main body of this thesis. The formation of these square sheets is usually associated with a more hypervalent (electron-rich) bonding arrangement in the square lattice where stabilization of the structure is achieved through additional interactions between the square nets and neighboring atoms [111].

Lastly, it is appropriate to touch upon intermetallic compounds containing three-dimensional anionic networks of p-block elements like that for the polymorphic RE_3Ge_5 (RE = Sm, Gd−Dy) [26, 27] or those of EuInGe and SrInGe [133] and Yb_2AuGe_3 [134], all of which have structures that are reminiscent of the substructure arrangements shown in Scheme 2.1. Other examples of networks can be shown with the high pressure forms of $LaGe_5$ [135] or $SrGe_{6-x}$ [136], the cluster networks in Eu_3Ge_5 [137], and even some clathrate-like Ge-networks, such as that in $Ba_{24}Ge_{100}$ [138]. There are several cases as well where the two-dimensional three-bonded layers which were found for the honeycomb-like moiety undergo a distortion or puckering resulting in the formation of fused six-membered Ge nets of which a single unit resembles the popular "chair" conformation adopted by cyclohexane. Some examples include the structures of

CaGe₂ [25], EuGe₂ [79], and the ternary compounds of EuGaGe [139] and isostructural EuGaSn. The change from planar to puckered nets has been demonstrated very nicely for the EuGaTt (Tt = Si, Ge, Sn) series [139], where the structure was shown to be dependent on the electronic factors and atomic sizes in the crystal structure.

Some other noteworthy germanium-containing compounds with unusual structural motifs or properties include:

(1) the Ge_2 dumbbells and 4-membered Ge chains in Sr_7Ge_6 and Ba_7Ge_6 that exhibit Ge π-bonding [140];

(2) the binary $EuGe_3$ compound, the structure of which is built up from triangular Ge_3 units, with the Eu cations ordering antiferromagnetically below 36 K [141];

(3) $RERh_4Ge_2$ (RE = Y, Gd–Lu) germanides in which unusual three-dimensional [Rh_4Ge_2] framework is found [142];

(4) The phase YFe_2Ge_2, crystallizing with the well-known $ThCr_2Si_2$-type structure, with interesting similarities suggested when comparing to the electronic structure of KFe_2As_2 (when the collapsed phase is formed at high pressure). This points at a similar pairing mechanism in place between the two compounds. Photoemission spectroscopy and heat capacity also show that YFe_2Ge_2 exhibits moderate heavy-fermion behavior with a Sommerfeld coefficient of 100 mJ/mol-K² [143, 144]

(5) UBeGe, a ternary phase with the ZrBeSi-type structure that contains BeGe honeycomb-like layers, with layers of U in between layer. The U atoms exhibit localized $5f$ states and a uniaxial ferromagnetic moment measured at $\mu_{\text{eff}} = 3.1\mu_B$ [145]

(6) $Pr_2Pt_3Ge_5$, which crystallizes with the orthorhombic $U_2Co_3Si_5$-type structure featuring three-dimensional framework of 3- and 4-coordinated Ge atoms [146]. The material is reported to have several low-temperature transitions one being superconductivity at 7.8 K and two successive antiferromagnetic transitions named as $T_{N1} = 3.5$ K and $T_{N2} = 4.2$ K, suggesting the long-range-ordered magnetism and superconductivity may coexist. Similar complex magnetic ordering has also been observed for the Pu "2–3–5" compounds [147]

(7) The $CeRuGe_5$ compound, which is seen as a derivative of the well-known tetragonal $HoCoGa_5$ structure type, shows Curie–Weiss paramagnetic behavior consistent with a Ce^{3+} ground state with no magnetic ordering detected down to 1.7 K [148]

Overall, the structural variety available for Ge seems to be not at all limited to a single form, and the illustrated compounds can be considered as representatives for the way in which structural chemistry can be understood and then manipulated for property optimizations or for application of the gained knowledge to other systems.

2.4. Heavy-Fermion Materials: The Ultimate Quantum Materials?

Particularly for the lanthanide or actinide binary and ternary compounds, which uniquely contain unpaired f-shell electrons, much work has been done in hopes to manipulate their physical properties through atomic substitutions, temperature, applied magnetic field, and pressure. Since the specific arrangements of atoms defining a crystal structure have a significant impact on how these properties are brought about (while taking into account the diverse arrangements already discussed), the theme of what function these f-electrons contribute to the properties has become a challenging area to understand. This particularly concerns finding the close relationship between the structural dimensionality and the electronic correlations that exist when coupled together.

At the forefront of f-electron quantum materials, nowadays one is sure to encounter heavy fermion (HF) compounds and their exciting physics [149–151]. These are a special class of materials in which there is a lattice of localized magnetic moments and conduction electrons with the f-states deriving very high effective masses. The term heavy fermion was applied to a class of intermetallic compounds that have electron density of states (DOS) many times larger than simple metals such as copper. The localized moments in these materials defined with the Ruderman–Kittel–Kasuya–Yoshida (RKKY) interaction and the Kondo effect will compete with one another. This is usually qualitatively represented in the well-known Doniach diagram. HF-compounds will have a lattice of magnetic moments that are assembled from lanthanide or actinide elements in intermetallics. These materials will reside at the boundary of electronic structure instability where the unpaired electrons are at the focus. Intermetallics containing Ce, Yb, U, and Pu are most-studied due to the $4f$- and $5f$-electronic ground states [152–158].

Particularly for the lanthanide or rare-earth metal binary and ternary compounds, which uniquely contain unpaired f-shell electrons, much work has been done in hopes to manipulate their physical properties through atomic substitutions, temperature, applied magnetic field, and pressure. This has been exemplified very nicely in some Ce- and Yb-based ternaries where these two lanthanide ions offer nearly empty or nearly filled $4f$ shells. Using chemical substitutions in $CeRhIn_5$, much like described in previous sections for unrelated compounds, the electronic states on both the transition metal or the Ce sites can be altered, which in turn can have noticeable effects on the long-range magnetic ordering. This is reported for $CeRh_{1-x}Ir_xIn_5$ ($x \leq 0.10$), where the Néel temperature decreases from the pure $CeRhIn_5$ to $Ce_{1-x}La_xRhIn_5$ ($x \leq 0.50$) [159]. Upon reaching the composition $Ce_{0.5}La_{0.5}RhIn_5$, T_N is so low that the material shows no magnetic order above 0.34 K.

Subjecting some intermetallics to liquid helium temperatures can show instances of superconductivity or magnetic order such as $CeCoIn_5$ [160] with a Curie temperature of 2.5 K and $CeRhIn_5$ $T_N = 3.8$ K and while applying pressure to the latter material superconductivity emerges at a Curie temperature of 2.1 K [161]. Lastly, magnetic fields can also be used to tune quantum materials and their electronic ground states such as the quantum critical point (QCP) in $YbRh_2Si_2$ [1562]. Since the specific arrangements of atoms defining a crystal structure have a significant impact on how these properties are brought about (while taking into account the diverse arrangements already discussed), the topic of what function these f-electrons contribute to the properties has become a challenging area to understand. Topology has also been shown to play an important role for intermetallic compounds in the realm of topological insulators [163, 164]. This is another emerging field which can lead to new materials with unique properties such as those reported for SmB_6 [165], PuB_6 [166], and PuB_4 [167].

Bibliography

[1] A. A. Gewirth, J. A. Varnell, and A. M. DiAscro, *Chem. Rev.*, **118**, 2313 (2018).

[2] Y. Xiong, L. Xiao, Y. Yang, F. J. DiSalvo, and H. D. Abruna, *Chem. Mater.*, **30**, 1532 (2018).

[3] The rise of quantum materials, *Nature Phys.*, **12**, 105 (2016). https://doi.org/10.1038/nphys3668

[4] J. E. Taylor, E. G. Teich, P. F. Damasceno, Y. Kallus, and M. Senechal, *Symmetry*, **9**, 188 (2017).

[5] M. W. Anderson, J. T. Gebbie-Rayet, A. R. Hill, N. Farida, M. P. Attfield, P. Cubillas, V. A. Blatov, D. M. Proserpio, D. Akporiaye, B. Arstad, and J. D. Gale, *Nature*, **544**, 456 (2017).

[6] D. J. Fredeman, P. H. Tobash, M. A. Torrez, J. D. Thompson, E. D. Bauer, F. Ronning, W. W. Tipton, S. P. Rudin, and R. G. Hennig, *Phys. Rev. B*, **83**, 224102 (2011).

[7] M. E. Jones, and R. E. Marsh, *J. Am. Chem. Soc.*, **76**, 1434 (1954).

[8] J. Nagamatsu, N. Nakagawa, T. Muranaka, Y. Zenitani, and J. Akimitsu, *Nature*, **410**, 6824 (2001).

[9] Y. Kong, O. V. Dolgov, O. Jepsen, and O. K. Andersen, *Phys. Rev. B*, **64**, 020501 (2001).

[10] M. Imai, A. Sato, T. Aoyagi, T. Kimura, Y. Matsushita, and N. Tsujii, *J. Am. Chem. Soc.*, **130**, 2886 (2008).

[11] P. Villars, and L. D. Calvert, *Pearson's Handbook of Crystallographic Data for Intermetallic Phases*, 2nd edn. (American Society for Metals, Materials Park, OH, 1991).

[12] N. Nagaosa, and Y. Tokura, *Nature Nanotechnol.*, **8**, 899 (2013).

[13] G. Finocchio, F. Büttner, R. Tomasello, M. Carpentieri, and M. Kläui, *J. Phys. D: Appl. Phys.*, **49**, 423001 (2016).

[14] M. Garst, J. Waizner, and D. Grundler, *J. Phys. D: Appl. Phys.*, **50**, 293002 (2017).

[15] Y. Ishikawa, K. Tajima, D. Bloch, and M. Roth, *Solid State Commun.*, **19**, 525 (1976).

[16] Y. Yufan, *Phys. Rev. Lett.*, **110**, 117202 (2013).

[17] B. Lebech, J. Bernhard, and T. Freltoft, *J. Phys. Condens. Matter*, **1**, 6105 (1989).

[18] N. Kanazawa, Y. Onose, T. Arima, D. Okuyama, K. Ohoyama, S. Wakimoto, K. Kakurai, S. Ishiwata, and Y. Tokura, *Phys. Rev. Lett.*, **106**, 156603 (2011).

[19] T. B. Massalski, *Binary Alloy Phase Diagrams* (American Society for Metals, Materials Park, OH, 1990).

[20] J. R. Salvador, J. R. Gour, D. Bilc, S. D. Mahanti, and M. G. Kanatzidis, *Inorg. Chem.*, **43**, 1403 (2004).

[21] J. R. Salvador, D. Bilc, S. D. Mahanti, and M. G. Kanatzidis, *Angew. Chem. Int. Ed.*, **42**, 1929 (2003).

[22] S. Nakatsuji, K. Kuga, Y. Machida, T. Tayama, T. Sakakibara, Y. Karaki, H. Ishimoto, S. Yonezawa, Y. Maeno, E. Pearson, G. G. Lonzarich, L. Balicas, H. Lee, and Z. Fisk, *Nature Phys.*, **4**, 603 (2008).

[23] R. T. Macaluso, S. Nakatsuji, K. Kuga, E. L. Thomas, Y. Machida, Y. Maeno, Z. Fisk, and J. Y. Chan, *Chem. Mater.*, **19**, 1918 (2007).

[24] E. L. Thomas, D. P. Gautreaux, H.-O. Lee, Z. Fisk, and J. Y. Chan, *Inorg. Chem.*, **46**, 3010 (2007).

[25] P. H. Tobash, and S. Bobev, *J. Solid State Chem.*, **180**, 1575 (2007).

[26] P. H. Tobash, D. Lins, S. Bobev, N. Hur, J. D. Thompson, and J. L. Sarrao, *Inorg. Chem.*, **45**, 7286 (2006).

[27] P. H. Tobash, S. Bobev, J. D. Thompson, and J. L. Sarrao, *J. Alloys Compd.*, **488**, 533 (2009).

[28] N.-T. Suen, and S. Bobev, *Z. Anorg. Allg. Chem.*, **640**, 805 (2014).

[29] G. Venturini, I. Ijjaali, and B. Malaman, *J. Alloys Compd.*, **284**, 262 (1999).

[30] G. Venturini, I. Ijjaali, and B. Malaman, *J. Alloys Compd.*, **285**, 194 (1999).

[31] A. M. Guloy, and J. D. Corbett, *Inorg. Chem.*, **30**, 4789 (1991).

[32] J. Zhang, P. H. Tobash, W. D. Pryz, D. J. Buttrey, N. Hur, J. D. Thompson, J. L. Sarrao, and S. Bobev, *Inorg. Chem.*, **52**, 953 (2013).

[33] P. Boulet, E. Colineau, F. Wastin, P. Javorsky, J. C. Griveau, J. Rebizant, G. R. Stewart, and E. D. Bauer, *Phys. Rev. B*, **72**, 064438 (2005).

[34] W. Choe, V. K. Pecharsky, A. O. Pecharsky, K. A. Gschneidner, V. G. Young, Jr., and G. J. Miller, *Phys. Rev. Lett.*, **84**, 4617 (2000).

[35] L. Morellon, P. A. Algarabel, M. R. Ibarra, J. Blasco, and B. García-Landa, *Phys. Rev. B*, **58**, R14721 (1998).

[36] V. K. Pecharsky, A. O. Pecharsky, and K. A. Gschneidner, Jr., *J. Alloys Compd.*, **344**, 362 (2002).

[37] Y. Mozharivskyj, A. O. Pecharsky, V. K. Pecharsky, and G. J. Miller, *J. Am. Chem. Soc.*, **127**, 317 (2005).

[38] E. M. Levin, K. A. Gschneidner, Jr., and V. K. Pecharsky, *Phys. Rev. B*, **65**, 214427 (2002).

[39] C. Magen, Z. Arnold, L. Morellon, Y. Skorokhod, P. A. Algarabel, M. R. Ibarra, and J. Kamarad, *Phys. Rev. Lett.*, **91**, 207202-1 (2003).

[40] K. A. Gschneidner, Jr., V. K. Pecharsky, and A. O. Tsokol, *Rep. Prog. Phys.*, **68**, 1479 (2005).

[41] E. M. Levin, V. K. Pecharsky, and K. A. Gschneidner, Jr., *Phys. Rev. B*, **60**, 7993 (1999).

[42] E. M. Levin, V. K. Pecharsky, and K. A. Gschneidner, Jr., *J. Magn. Magn. Matter.*, **210**, 181 (2000).

[43] E. M. Levin, V. K. Pecharsky, and K. A. Gschneidner, Jr., *Phys. Rev. B*, **63**, 174110 (2001).

[44] L. Morellon, P. A. Algarabel, C. Magen, and M. R. Ibarra, *J. Magn. Magn. Mater.*, **237**, 119 (2001).

[45] L. Morellon, J. Blasco, P. A. Algarabel, and M. R. Ibara, *Phys. Rev. B*, **62**, 1022 (2000).

[46] C. Magen, L. Morellon, P. A. Algarabel, C. Marquina, and M. R. Ibara, *J. Phys.: Condens. Matter*, **15**, 2389 (2003).

[47] L. Morellon, J. Stankiewicz, B. Garcia-Landa, P. A. Algarabel, and M. R. Ibara, *Appl. Phys. Lett.*, **73**, 3462 (1998).

[48] W. Choe, A. O. Pecharsky, M. Wrle, and G. J. Miller, *Inorg. Chem.*, **42**, 8223 (2003).

[49] G. J. Miller, *Chem. Soc. Rev.*, **35**, 799 (2006).

[50] A. O. Pecharsky, K. A. Gschneidner, Jr., V. K. Pecharsky, and C. E. Schindler, *J. Alloys Compd.*, **338**, 126 (2002).

[51] V. K. Pecharsky, G. D. Samolyuk, V. P. Antropov, A. O. Pecharsky, and K. A. Gschneidner, Jr., *J. Solid State Chem.*, **171**, 57 (2003).

[52] A. O. Pecharsky, K. A. Gschneidner, Jr., V. K. Pecharsky, D. L. Schlagel, and T. A. Lograsso, *Phys. Rev. B*, **70**, 144419 (2004).

[53] A. M. Pereira, J. B. Sousa, J. P. Araujo, C. Magen, P. A. Algarabel, L. Morellon, C. Marquina, and M. R. Ibarra, *Phys. Rev. B*, **77**, 134404 (2008).

[54] K. Ahn, A. O. Tsokol, Yu. Mozharivskyj, K. A. Gschneidner, Jr., and V. K. Pecharsky, *Phys. Rev. B*, **72**, 054404 (2005).

[55] C. Ritter, L. Morellon, P. A. Algarabel, C. Magen, and M. R. Ibarra, *Phys. Rev. B*, **65**, 094405 (2002).

[56] M. Zou, V. K. Pecharsky, K. A. Gschneidner, Jr., D. L. Schlagel, and T. A. Lograsso, *Phys. Rev. B*, **78**, 014435 (2008).

[57] V. K. Pecharsky, and K. A. Gschneidner, Jr., *Phys. Rev. Lett.*, **78**, 4494 (1997).

[58] M. Zou, V. K. Pecharsky, K. A. Gschneidner, Ya. Mudryk, D. L. Schlagel, and T. A. Lograsso, *Phys. Rev. B: Codens. Matter Mater. Phys.*, **80**, 174411 (2009).

[59] D. M. Proserpio, G. Chacon, and C. Zheng, *Chem. Mater.*, **10**, 1286 (1998).

[60] C. Zheng, R. Hoffmann, R. Nesper, and H.-G. von Schnering, *J. Am. Chem. Soc.*, **108**, 1876 (1986).

[61] C. Zheng, and R. Hoffmann, *J. Am. Chem. Soc.*, **108**, 3078 (1986).

[62] U. Häussermann, S. Amerioun, L. Erikson, C.-S. Lee, and G. J. Miller, *J. Am. Chem. Soc.*, **124**, 4371 (2002).

[63] G. Just, and P. Paufler, *J. Alloys Compd.*, **232**, 1 (1996).

[64] G. J. Miller, *Eur. J. Inorg. Chem.*, 523 (1998).

[65] S. Ponou, and T. F. Fässler, *Inorg. Chem.*, **43**, 6124 (2004).

[66] T. Kyratsi, D. Y. Chung, and M. G. Kanatzidis, *J. Alloys Compd.*, **338**, 36 (2002).

[67] Y. Mozharivskyj, D. Kaczorowski, and H. F. Franzen, *J. Solid State Chem.*, **155**, 259 (2000).

[68] N.-T. Suen, J. Hooper, E. Zurek, and S. Bobev, *J. Am. Chem. Soc.*, **134**, 12708 (2012).

[69] P. H. Tobash, J. J. Meyers, G. DiFilippo, S. Bobev, F. Ronning, J. D. Thompson, and J. L. Sarrao, *Chem. Mater.*, **20**, 2151 (2008).

[70] P. H. Tobash, S. Bobev, F. Ronning, J. D. Thompson, and J. L. Sarrao, *J. Alloys Compd.*, **488**, 511 (2009).

[71] M. Tanaka, H. Horiuchi, T. Shishido, and T. Fukuda, *Acta Cryst.*, **C49**, 437 (1993).

[72] S.-J. Kim, S. Hu, C. Uher, T. Hogan, B. Huang, J. D. Corbett, and M. G. Kanatzidis, *J. Solid State Chem.*, **153**, 321 (2000).

[73] A. K. Ganguli, and J. D. Corbett, *J. Solid State Chem.*, **107**, 480 (1993).

[74] S. M. Kauzlarich, (ed.), *Chemistry, Structure and Bonding of Zintl Phases and Ions*; (VCH Publishers, New York, 1996), and the references therein.

[75] J.-T. Zhao, and J. D. Corbett, *Inorg. Chem.*, **34**, 378 (1995).

[76] J. Y. Chan, M. E. Wang, A. Rehr, D. J. Webb, and S. M. Kauzlarich, *Chem. Mater.*, **9**, 2131 (1997).

[77] F. Zurcher, and R. Nesper, *Angew. Chem. Int. Ed.*, **37**, 3314 (1998).

[78] Z. Xu, and A. M. Guloy, *J. Am. Chem. Soc.*, **120**, 7349 (1998).

[79] S. Bobev, V. Fritsch, J. D. Thompson, J. L. Sarrao, B. Eck., R. Dronskowski, and S. M. Kauzlarich, *J. Solid State Chem.*, **178**, 1071 (2005).

[80] S. Ponou, T. F. Fässler, G. Tobías, E. Canadell, A. Cho, and S. C. Sevov, *Chem. Eur. J.*, **10**, 3615 (2004).

[81] S.-Q. Xia, J. Hullmann, S. Bobev, A. Ozbay, and E. R. Nowak, *J. Solid State Chem.*, **180**, 2088 (2007).

[82] A. M. Mills, R. Lam, M. J. Ferguson, L. Deakin, and A. Mar, *Coord. Chem. Rev.*, **233**, 207 (2002), and the references therein

[83] Z. Xu, and A. M. Guloy, *J. Am. Chem. Soc.*, **119**, 10541 (1997).

[84] A. M. Mills, and A. Mar, *Inorg. Chem.*, **39**, 4599 (2000).

[85] J. R. Salvador, D. Bilc, S. D. Mahanti, T. Hogan, F. Guo, and M. G. Kanatzidis, *J. Am. Chem. Soc.*, **126**, 4474 (2004).

[86] Q. Xie, and R. Nesper, *Z. Anorg. Allg. Chem.*, **632**, 1743 (2006).

[87] J.-G. Mao, Z. Xy, and A. M. Guloy, *Inorg. Chem.*, **40**, 4472 (2001).

[88] A. M. Mills, and A. Mar, *J. Am. Chem. Soc.*, **123**, 1151 (2001).

[89] J. D. Bryan, and G. D. Stucky, *Chem. Mater.*, **13**, 253 (2001).

[90] D.-K. Seo, and J. D. Corbett, *J. Am. Chem. Soc.*, **123**, 4512 (2001).

[91] S.-J. Kim, and M. G. Kanatzidis, *Inorg. Chem.*, **40**, 3781 (2001).

[92] J.-G. Mao, J. Goodey, and A. M. Guloy, *Inorg. Chem.*, **41**, 931 (2002).

[93] J. Jiang, and S. M. Kauzlarich, *Chem. Mater.*, **18**, 435 (2006).

[94] A. V. Tkachuk, and A. Mar, *J. Solid State Chem.*, **180**, 2298 (2007).

[95] B. Li, and J. D. Corbett, *Inorg. Chem.*, **46**, 2237 (2007).

[96] P. Alemany, M. Llunell, and E. Canadell, *Inorg. Chem.*, **45**, 7235 (2006).

[97] D.-K. Seo, and J. D. Corbett, *J. Am. Chem. Soc.*, **122**, 9621 (2000).

[98] G. A. Papoian, and R. Hoffmann, *J. Am. Chem. Soc.*, **123**, 6600 (2001).

[99] M. L. Munzarova, and R. Hoffmann, *J. Am. Chem. Soc.*, **124**, 4787 (2002).

[100] F. J. DiSalvo, *Pure Appl. Chem.*, **72**, 1799 (2000).

[101] M. Jansen, *Angew. Chem. Int. Ed.*, **41**, 3746 (2002).

[102] F. Weitzer, K. Hiebl, and P. Rogl, *J. Solid State Chem.*, **98**, 291 (1992).

[103] J. X. Boucherle, F. Givord, P. Lejay, J. Schweizer, and A. Stunault, *Acta Cryst.*, **B44**, 377 (1988).

[104] E. D. Bauer, P. H. Tobash, J. N. Mitchell, and J. L. Sarrao, *Phil. Mag.*, **92**, 2466 (2012).

[105] J. L. Sarrao, L. A. Morales, J. D. Thompson, B. L. Scott, G. R. Stewart, F. Wastin, J. Rebizant, P. Boulet, E. Colineau, and G. H. Lander, *Nature*, **420**, 297 (2002).

[106] H. B. Rhee, F. Ronning, J.-X. Zhu, E. D. Bauer, J. N. Mitchell, P. H. Tobash, B. L. Scott, J. D. Thompson, Y. Jiang, C. H. Booth, and W. E. Pickett, *Phys. Rev. B*, **86**, 115137 (2012).

[107] T. G. Akhmetshina, V. A. Blatov, D. M. Proserpio, and A. P. Shevchenko, *Acc. Chem. Res.*, **51**, 21 (2018).

[108] A. O. Oliynyk, and A. Mar, *Acc. Chem. Res.*, **51**, 59 (2018).

[109] S. K. McCall, M. F. Fluss, B. W. Chung, M. W. McElfresh, D. D. Jackson, and G. F. Chapline, *Proc. Natl. Acad. Sci. USA*, **103**, 17179 (2006).

[110] C. H. Booth, Y. Jiang, S. A. Medling, D. L. Wang, A. L. Costello, D. S. Schwartz, J. N. Mitchell, P. H. Tobash, E. D. Bauer, S. K. McCall, M. A. Wall, and P. G. Allen, *J. App. Phy.*, **113**, 093502 (2013).

[111] G. A. Papoian, and R. Hoffmann, *Angew. Chem. Int. Ed.*, **39**, 2409 (2000).

[112] P. Schobinger-Papamantellos, and K. H. J. Buschow, *J. Magn. Magn. Mater.*, **82**, 99 (1989).

[113] P. Schobinger-Papamantellos, and K. H. J. Buschow, *J. Less-Common Met.*, **146**, 279 (1989).

[114] P. Schobinger-Papamantellos, and K. H. J. Buschow, *J. Less-Common Met.*, **163**, 319 (1990).

[115] O. Zaharko, P. Schobinger-Papamantellos, and C. Ritter, *J. Alloys Compd.*, **280**, 4 (1998).

[116] J. Mao, and A. M. Guloy, *J. Alloys Compd.*, **322**, 135 (2001).

[117] G. S. Smith, Q. Johnson, and A. G. Tharp, *Acta Cryst.*, **23**, 640 (1967).

[118] L. Pauling, *The Nature of the Chemical Bond*, 3rd edn. (Cornell University Press, Ithaca, NY, 1960).

[119] A. M. Guloy, and J. D. Corbett, *Inorg. Chem.*, **35**, 2616 (1996).

[120] A. V. Mudring, and J. D. Corbett, *J. Am. Chem. Soc.*, **126**, 5277 (2004).

[121] P. H. Tobash, D. Lins, S. Bobev, A. Lima, M. F. Hundley, J. D. Thompson, and J. L. Sarrao, *Chem. Mater.*, **17**, 5567 (2005).

[122] N.-T. Suen, P. H. Tobash, and S. Bobev, *J. Solid State Chem.*, **184**, 2941 (2011).

[123] P. H. Tobash, and S. Bobev, *J. Am. Chem. Soc.*, **128**, 3532 (2006).

[124] P. H. Tobash, S. Bobev, J. D. Thompson, and J. L. Sarrao, *Inorg. Chem.*, **48**, 6641 (2009).

[125] R. Kraft, and R. Pöttgen, *Montash. Chem.*, **135**, 1327 (2004).

[126] W. Choe, G. Miller, and E. Levin, *J. Alloys Compd.*, **329**, 121 (2001).

[127] P. H. Tobash, G. DiFilippo, S. Bobev, N. Hur, J. D. Thompson, and J. L. Sarrao, *Inorg. Chem.*, **46**, 8690 (2007).

[128] P. S. Salamakha, O. L. Sologub and O. I. Bodak, in K. A. Gschneidner, Jr., and L. Eyring, (eds.), *Handbook on the Physics and Chemistry of Rare Earths*, Vol. 27 (North-Holland, Amsterdam, 1999), p. 1.

[129] G. Venturini, I. Ijjaali, and B. Malaman, *J. Alloys Compd.*, **289**, 168 (1999).

[130] A. Grytsiv, D. Kaczorowski, A. Leithe-Jasper, P. Rogl, M. Potel, H. Noël, A. P. Pikul, and T. Velikanova, *J. Solid State Chem.*, **165**, 178 (2002).

[131] G. Venturini, *J. Alloys Compd.*, **308**, 200 (2000).

[132] G. Venturini, I. Ijjaali, and B. Malaman, *J. Alloys Compd.*, **288**, 183 (1999).

[133] J.-G. Mao, J. Goodey, and A. M. Guloy, *Inorg. Chem.*, **41**, 931 (2002).

[134] S. Sarkar, and S. C. Peter *J. Chem. Sci.*, **124**, 1385 (2012).

[135] H. Fukuoka, and S. Yamanaka, *Phys. Rev. B*, **67**, 094501 (2003).

[136] H. Fukuoka, S. Yamanaka, E. Matsuoka, and T. Takabatake, *Inorg. Chem.*, **44**, 1460 (2005).

[137] S. Budnyk, F. Weitzer, C. Kubata, Y. Prots, L. G. Akselrud, W. Schnelle, K. Hiebl, R. Nesper, F. R. Wagner, and Y. Grin, *J. Solid State Chem.*, **179**, 2329 (2006).

[138] H. Fukuoka, K. Iwai, S. Yamanaka, H. Abe, K. Yoza, and L. Hming, *J. Solid State Chem.*, **151**, 117 (2000).

[139] T.-S. You, Y. Grin, and G. J. Miller, *Inorg. Chem.*, **46**, 8801 (2007).

[140] L. Siggelkow, V. Hlukhyy, and T. F. Fässler, *J. Solid State Chem.*, **191**, 76 (2012).

[141] R. Castillo, A. I. Baranov, U. Burkhardt, Y. Grin, and U. Schwarz, *Z. Anorg. Allg. Chem.*, **641**, 355 (2015).

[142] D. VoBwinkel, S. F. Matar, and R. Pöttgen, *Monatsh Chem.*, **146**, 1375 (2015).

[143] J. Chen, K. Semeniuk, Z. Feng, P. Reiss, P. Brown, Y. Zou, P. W. Logg, G. I. Lampronti, and F. M. Grosche, *Phys. Rev. Lett.*, **116**, 127001 (2016).

[144] D. F. Xu, D. W. Shen, J. Jiang, B. P. Xie, Q. S. Wang, B. Y. Pan, P. Dudin, T. K. Kim, M. Hoesch, J. Zhao, X. G. Wan, and D. L. Feng, *Phys. Rev. B*, **93**, 024506 (2016).

[145] R. Gumeniuk, A. N. Yaresko, W. Schnelle, M. Nicklas, K. O. Kvasnina, C. Hennig, Y. Grin, and A. Leithe-Jasper, *Phys. Rev. B*, **97**, 174405 (2018).

[146] N. H. Sung, C. J. Roh, K. S. Kim, and B. K. Cho, *Phys. Rev. B*, **86**, 224507 (2012).

[147] E. D. Bauer, P. H. Tobash, J. N. Mitchell, J. A. Kennison, F. Ronning, B. L. Scott, and J. D. Thompson, *J. Phys.: Condens. Matter*, **23**, 094223 (2011).

[148] E. Murashova, Zh. Kurenbaeva, A. Gribanov, and D. Kaczorowski, *J. Alloys Compds.*, **701**, 626 (2017).

[149] Q. Si, and F. Steglich, *Science*, **329**, 1161 (2010).

[150] P. Coleman, arXiv:0612006v3.

[151] P. Gegenwart, Q. Si, and F. Steglich, *Nature Phys.*, **4**, 186 (2008).

[152] J. L. Sarrao, E. D. Bauer, J. N. Mitchell, P. H. Tobash, and J. D. Thompson, *Physica C*, **514**, 184 (2015).

[153] R. Settai, T. Takeuchi, and Y. Onuki, *J. Phys. Soc. Japan*, **76**, 051003 (2007).

[154] P. H. Tobash, Y. Jiang, F. Ronning, C. H. Booth, J. D. Thompson, B. L. Scott, and E. D. Bauer, *J. Phys.: Codens. Matter*, **23**, 086002 (2011).

[155] P. H. Tobash, F. Ronning, J. D. Thompson, S. Bobev, and E. D. Bauer, *J. Solid State Chem.*, **183**, 707 (2010).

[156] J. Flouquet, and H. Harima, arXiv:0910.3110.

[157] Y. Onuki, F. Honda, Y. Hirose, R. Settai, and T. Takeuchi, *Rep. Prog. Phys.*, **79**, 114501 (2016).

[158] A. K. Pathak, D. Paudyal, Y. Mudryk, and V. K. Pecharsky, *Phys. Rev. B*, **96**, 064412 (2017).

[159] B. E. Light, R. S. Kumar, A. L. Cornelius, P. G. Pagliuso, and J. L. Sarrao, *Phys. Rev. B*, **69**, 024419-1 (2004).

[160] J. Paglione, M. A. Tanatar, D. G. Hawthorn, E. Boaknin, R. W. Hill, F. Ronning, M. Sutherland, L. Taillefer, C. Petrovic, and P. C. Canfield, *Phys. Rev. Lett.*, **91**, 246405 (2003).

[161] H. Hegger, C. Petrovic, E. G. Moshopoulou, M. F. Hundley, J. L. Sarrao, Z. Fisk, and J. D. Thompson, *Phys. Rev. Lett.*, **84**, 4986 (2000).

[162] P. Gegenwart, J. Custers, C. Geibel, K. Neumaier, T. Tayama, K. Tenya, O. Troarelli, and F. Steglich, *Phys. Rev. Lett.*, **89**, 056402-1 (2002).

[163] M. Z. Hasan, and C. L. Kane, *Rev. Mod. Phys.*, **82**, 3045 (2010).

[164] X. L. Qi, and S. C. Zhang, *Rev. Mod. Phys.*, **83**, 1057 (2011).

[165] M. Dzero, K. Sun, V. Galitski, and P. Coleman, *Phys. Rev. Lett.*, **104**, 106408 (2010).

[166] X. Deng, K. Haule, and G. Kotliar, *Phys. Rev. Lett.*, **111**, 176404 (2013).

[167] H. Choi, W. Zhu, S. K. Cary, L. E. Winter, Z. Huang, R. D. McDonald, V. Mocko, B. L. Scott, P. H. Tobash, J. D. Thompson, S. A. Kozimor, E. D. Bauer, J.-X. Zhu, and F. Ronning, *Phys. Rev. B*, **97**, 201114 (2018).

Chapter 3

Solution Growth of Intermetallic Single Crystals

Raquel A. Ribeiro[*,†] and Paul C. Canfield[†,‡]

*CCNH, Universidade Federal do ABC (UFABC),
Santo André, SP, Brazil
†Department of Physics and Astronomy, Iowa State University,
Ames, IA, USA
‡Ames Laboratory, Iowa State University, Ames, IA, USA

3.1. Think–Make–Measure–Think

The field of New Materials Physics is driven by the design, discovery and growth of novel materials that are then characterized, understood and ultimately mastered. At its simplest level, the flowchart for research in this field is a repeating loop of think–make–measure–think. In this chapter, we want to provide an introduction to solution growth of single crystals; one of the most versatile techniques available for the "make" step of this loop. In order to give at least a small flavor or the initial "think" step, we will provide some rational/back-story for the two compounds we will use as examples for this chapter: $PtSn_4$ and $BaBi_3$.

The understanding of the physical properties of topological insulators has been a very active field in recent years. These materials are strong candidates for applications in spintronics, thanks to their unusual electronic properties [1]. After the discovery of quantum Hall states in quantum wells of HgTe [2], it was predicted that Bi_2Se_3, Bi_2Te_3 and Sb_2Te_3 could also have topologically protected surface states [3–6]. The bulk states of these compounds have a gap, but they have a single Dirac cone on the surface that represents the topologically protected surface states. In addition to such topological insulators, Dirac and Weyl semimetals can have topologically non-trivial states and potentially manifest exotic electronic properties [7, 8] including exceptionally large magnetoresistance (MR). In general, the likelihood of finding such states increases with the degree of spin–orbit–coupling (SOC) associated with the conduction electrons. At the simplest level, SOC increases with the number of protons in the nuclei of the constituent elements. This helps explaining the predominance of heavier elements in the compounds mentioned above.

$PtSn_4$ is a compound that was one of the earliest examples of exceptionally large MR. It has a ~500,000% MR at 1.8 K and 140 kOe [9]. Based on its phenomenal MR, as well as the presence element of fifth period Sn and sixth period element Pt (i.e., heavy elements), $PtSn_4$ was studied in detail by angle-resolved photoemission spectroscopy (ARPES). As such, $PtSn_4$ ended up being one of the rare, experimentally discovered topological semimetals, manifesting Dirac node arcs [10]. One of the other advantages of $PtSn_4$ was the fact that it could be readily grown out of excess Sn at very modest temperatures. This compound is an ideal practice/test system for any beginning crystal growth given that Pt and Sn are stable and non-toxic and the temperature range for the growth is relatively low.

$BaBi_3$ offers a different set of tantalizing features. First of all, it is exceptionally rich in very heavy elements with both Ba and Bi coming from the sixth period and Bi being, arguably, the heaviest stable element known to date.[1] Secondly, $BaBi_3$ is a superconductor, offering the possibility of combining topological effects with superconductivity. Thirdly, there were indications that $BaBi_3$ may have a rich pressure–temperature phase diagram, with multiple superconducting phases existing at low temperatures [11]. All of these features combined to offer the possibility of finding some

---

[1]Bi has a half-life longer than age of universe, 1.9×10^{19} years.

mixture of superconductivity and novel electronic topology at low temperatures. This possibility made the added difficulty of working with an air-sensitive (and somewhat toxic) element like Ba worth the extra effort [12].

Then, $PtSn_4$ and $BaBi_3$, represent a pair of materials that are each rich in heavy elements and, each in its own way, offer the possibility of novel electronic topologically non-trivial states. They also offer a good pedagogical set of relatively simple binary growths. We will start then with $PtSn_4$ as an example of a binary growth using non-toxic, non-air-sensitive elements that can be handled in a rather casual manner. We will demonstrate how to progress from using a binary phase diagram to planning the growth to the experimental implementation of actually assembling such a growth. With the basics taken care of, we will then outline what modifications need to be made to grow an air-sensitive sample such as $BaBi_3$. These are the same modifications we use for growths of Co-doped $BaFe_2As_2$ [13] and pure and doped $CaKFe_4As_4$ [14]. These two examples effectively cover the techniques needed for a majority of growths of intermetallic materials and certainly offer an excellent introduction to solution growth of binary compounds.

3.2. Binary Phase Diagrams and How to Use Them

Now that we have our idea for target growths, i.e., $PtSn_4$ and $BaBi_3$, it is time to see what we can learn from the existing binary phase diagrams. This is akin to checking maps before heading out on a cross country drive. It is good to know what the basic lay of the land is: where there is land, lake, river or swamp. Binary phase diagrams serve the same purpose. They can tell us where there is liquid, where there is solid, where there are two-phase regions, or even three-phase points.

To start, let us examine the Sn-rich part of the Pt–Sn binary phase diagram shown in Figure 3.1. At high enough temperatures, there is a single-phase, liquid region (denoted by L in Figure 3.1); this is above 1268°C for $Pt_{50}Sn_{50}$ and as low as 231°C for pure Sn. The low temperature extent of this liquid region is called the liquidus line

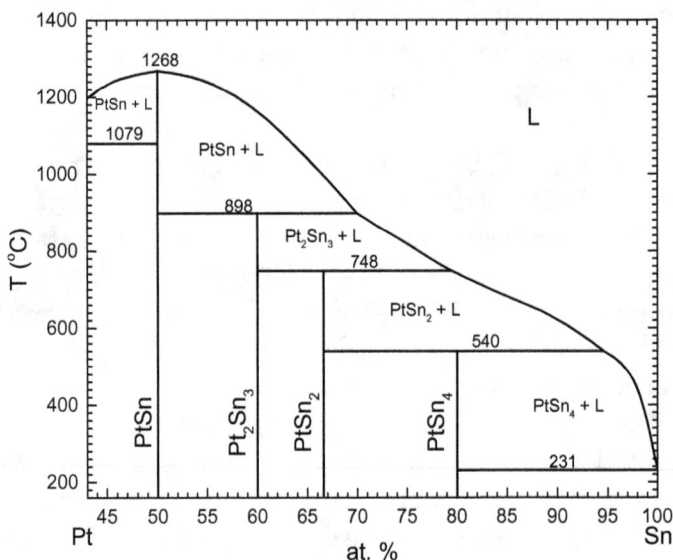

Figure 3.1.   Sn-rich part of the Pt–Sn binary phase diagram [15].

and forms the cascade of water-slide-like-curves from $Pt_{50}Sn_{50}$ down to Pt_0Sn_{100}. There is a discontinuity in this liquidus line each time either (1) a horizontal line (called a peritectic line) or (2) a vertical line (associated with a congruently melting compound) intersects it. In general, the liquidus line separates the single-phase (liquid) part of the binary phase diagram from two-phase regions. This may sound very formal, but it is the heart and essence of solution growth. We will come back to this in a few paragraphs, after we become familiar with the other parts of the binary phase diagram.

In addition to the high temperature, single-phase liquid region of the phase diagram, there are lower temperature features. There are four binary compounds (line compounds with no apparent widths of formation) and pure, elemental Sn (on the far right). Of these compounds, only PtSn extends all the way to the liquidus line. PtSn is a congruently melting compound. This means that at 1268°C it melts and forms a liquid with composition $Pt_{50}Sn_{50}$. This may sound trivial and common place because this is more or less what every day experience leads us to expect from melting of water. It is actually more the exception than the rule in binary phase diagrams. Note that Pt_2Sn_3, $PtSn_2$, and $PtSn_4$ do not extend all the way to the liquidus, instead they are truncated by horizontal lines associated with the peritectic decomposition of each of these solids when they are heated. We can take Pt_2Sn_3 as an example. When a lump of Pt_2Sn_3 is heated, it remains single phase until it reaches 898°C. If it is heated to a few degrees above 898°C it, decomposes into a liquid (with a composition of roughly $Pt_{30}Sn_{70}$) and PtSn. As temperature is increased, there will be less and less PtSn and the liquid will approach a $Pt_{40}Sn_{60}$ composition. Once the temperature increases past the value of the liquidus line for $Pt_{40}Sn_{60}$ (roughly 1160°C), the sample, which at low temperatures was our lump of Pt_2Sn_3, will be a single-phase liquid with composition $Pt_{40}Sn_{60}$.

In the previous paragraph, we examined what happened to PtSn and Pt_2Sn_3 when we warmed them up. Now let us see what happens when we cool down liquids of $Pt_{50}Sn_{50}$ and $Pt_{40}Sn_{60}$. This is when the difference between congruently melting and peritectically decomposing compounds becomes even more important. If we cool a liquid of $Pt_{50}Sn_{50}$ from 1300°C down to room temperature, it will change from a single-phase liquid to a single-phase solid: PtSn. If we take no effort to control nucleation, the PtSn will be polycrystalline with the grain size giving us some hint about how quickly we cooled through 1268°C. If we now cool a liquid of $Pt_{40}Sn_{60}$ over the same temperature range, we end up with something quite different. As we cool through ~1160°C we start to grow PtSn and have the remaining liquid become more Sn rich. We will continue to grow PtSn until we cool below 898°C. At that point we will start to grow Pt_2Sn_3, still having the remaining liquid becoming more and more Sn rich. As we cool below 748°C, we will start to grow $PtSn_2$ and once we cool below 540°C we will grow $PtSn_4$. So, instead of getting a single-phase Pt_2Sn_3 solid we end up with four phases: PtSn, Pt_2Sn_3, $PtSn_2$ and $PtSn_4$. Yikes! What a mess!

If we want to grow single crystal, single-phase Pt_2Sn_3 it can be readily done, but not by cooling a $Pt_{40}Sn_{60}$ liquid. Pt_2Sn_3 can be grown by cooling a liquid with a composition between $Pt_{29}Sn_{71}$ and $Pt_{21}Sn_{79}$ from above the liquidus line down to 748°C. For example, let our growth start with a liquid of composition $Pt_{28}Sn_{72}$ that will be cooled from 950°C (well within the single-phase liquid range) down to 760°C (well above the peritectic temperature for $PtSn_2$). As the temperature is lowered from 950°C, our growth stops being a single-phase, $Pt_{28}Sn_{72}$ liquid when its temperature reaches the liquidus line (near 880°C). At this point, crystals of Pt_2Sn_3 start to grow. Pt_2Sn_3 will continue to form as we continue to cool our growth. Since we want single-phase Pt_2Sn_3, we must stop cooling at a temperature above 748°C. At this point, we need to remove the Pt_2Sn_3 crystals from the remaining liquid, or, perhaps remove the remaining liquid from around the Pt_2Sn_3 crystals. The experimental manifestation of the prior sentence will be covered in detail in Section 3.3.

Having reviewed the basic features and implications of the Pt–Sn phase diagram, it should be rather obvious how to grow our target compound: $PtSn_4$. On the one hand, $PtSn_4$ has a peritectic decomposition at 540°C, so it cannot be readily grown by cooling a $Pt_{20}Sn_{80}$ melt. On the other hand, $PtSn_4$ has a clear and accessible liquidus line. When we first wanted to try growing $PtSn_4$, we cooled a melt of $Pt_{04}Sn_{96}$ from 600°C down to 350°C over 60 hours. This allowed for the formation of plate-like crystals with areas greater than 5 mm × 5 mm and thicknesses of up to 1 mm [9]. With samples like these detailed thermodynamic, transport and spectroscopic measurement could be made and $PtSn_4$ could be identified as an intermetallic with exceptionally large MR as well as Dirac node arcs [9, 10].

The second compound that we have identified for growth is $BaBi_3$; the Bi-rich part of the Ba–Bi binary phase diagram is shown in Figure 3.2. This phase diagram has two features that were not present in the Pt–Sn binary phase diagram shown in Figure 3.1. First, there is a clear eutectic point near $Ba_{02}Bi_{98}$ with an eutectic temperature of 262°C, clearly suppressed from the 271°C melting point of pure Bi. At the eutectic point three phases: $BaBi_3$, Bi, and the liquid phase co-exist. In theory, a similar type of eutectic must exist between $PtSn_4$ and pure Sn. The fact that it is not shown implies that it is very close to pure Sn and has not been experimentally detected. The second new feature in Figure 3.2 is the dashed line at the upper left

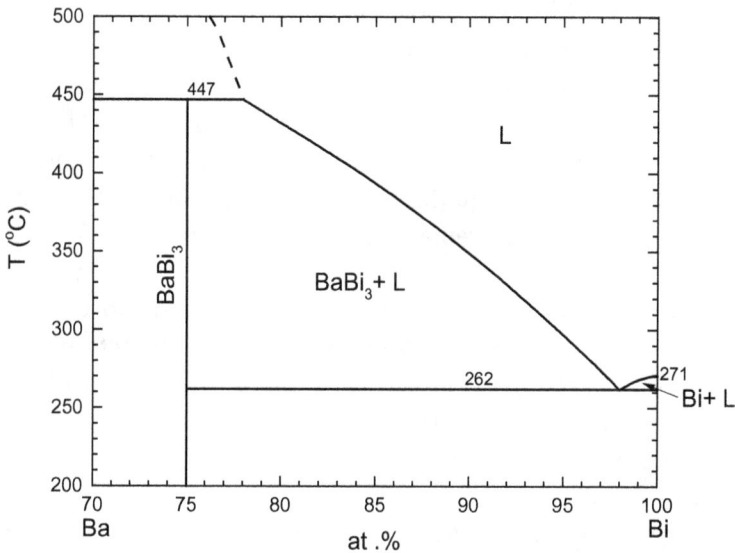

Figure 3.2. Bi-rich part of the Ba–Bi binary phase diagram [15].

of the figure representing the current, best guess, that there is some other Ba–Bi binary phase that forms between pure Ba and BaBi$_3$. The fact that this is a "best guess" is flagged by the dashed line.

It is worth pointing out that any phase diagram is merely a suggestion of reality, our current best guess at what actually happens. Many binary phase diagrams have missing parts or are not determined at all. When you get to ternary or higher phase diagrams (a topic beyond the scope of this limited review), the vast majority of diagrams are simply not explored or mapped at all. Even phase diagrams that are presented as "known" are revised on a regular basis. For example, over the past decade the Sc–Zn and Gd–Cd binary phase diagrams have been redefined by the[16, 17] discoveries of new, binary quasicrystalline phases in regions that were thought to be well understood and associated with ScZn$_6$ or GdCd$_6$ crystalline phases. Often the best way to actually determine points on a phase diagram is by making a growth and analyzing the resulting data [18].

Conceptually, the growth of BaBi$_3$ is not different from the growth of PtSn$_4$; we find the exposed liquidus line for BaBi$_3$ and cool through it. The details of how we change from elements that are non-reactive with air (Pt and Sn) to an element that needs to be handled in a non-reactive environment (Ba) will be discussed in detail in Section 3.3. A solution of $Ba_{20}Bi_{80}$ can be heated 600°C (well into the single-phase liquid region) and then cooled to 350°C (below the liquidus line) over 125 hours [11, 12, 19]. Again, at this point the remaining liquid and the crystals of $BaBi_3$ need to be separated from each other (see below). With single crystals of $BaBi_3$ in hand, thermodynamic and transport properties [11, 12, 19] can be measured and the detailed interaction between the structural and superconducting phases as a function of applied pressure can be determined [11, 12]. In essence, with $BaBi_3$ we start with a composition–temperature phase diagram and end up with a pressure–temperature phase diagram [12]. This use of one phase diagram to allow for the growth of a material that then allows for the determination of another type of phase diagram is at the heart of new materials physics.

Before focusing on the experimental reality of growing these samples, it is important to warn the reader that the above has been a cursory and rather functional review of some of the features of binary phase diagrams. For further details, a rather extensive and diverse collection of growth papers was assembled in a special issue of *Philosophical Magazine*, a "Symposium on Design, Discovery, and Growth of Novel Materials" [20]. In this issue, there are many articles with details about growth, both theoretical as well as exceptionally practical. The article by Fisher, Shapiro and Analytics is particularly helpful in understanding the theoretical basis of binary phase diagrams [21]. For details about ternary or quaternary growths, Ref. [22] provides an overview of several growths; Ref. [14] provides an exceptionally detailed glimpse of what the growth of a quaternary compound entails and illustrates how solution growth can be used to determine parts of complex, compositional–temperature phase diagrams.

But enough about such abstractions; on to the experimental details.

3.3. Experimental Details of Assembly and Growth of PtSn$_4$ and BaBi$_3$

3.3.1. *PtSn$_4$: Non-air-sensitive growth*

The first stage of growth planning begins with determining the concentration of the reagents that will be used to create the single-phase liquid at high temperatures. As we discussed in Section 3.2, and shown in Figure 3.1, the compositions that have the coexistence of PtSn$_4$ and liquid start approximately at Pt$_{05}$Sn$_{95}$ and go to almost Pt$_0$Sn$_{100}$. Since phase diagrams are, at best, approximations of reality and can have errors or uncertainties, we do not recommend starting with a concentration exactly where the liquidus line of the compound is purported to start, but rather slightly removed from this point, a bit further down the curve. In this case, we will use Pt$_{04}$Sn$_{96}$ as an initial composition. This helps to avoid the growth of another phase, which in this case would be PtSn$_2$.

Figures 3.3(a)–(c) show the steps associated with filling the growth crucible. For many growths, and particularly a first attempt to grow any specific phase, it is recommend starting with a lower volume crucible, e.g., a 2 ml rather than a 5 ml crucible. For this growth, we will use a 2 ml crucible set made out of Al$_2$O$_3$; the set (Figure 3.3(a)) consists of two crucibles and a frit and is referred to as a Canfield Crucible Set (CCS) [18]. As a general rule, it is good to plan to fill the crucible with reagents between 2/3 to 3/4 of its volume. Since some reagents are sold in large or small pieces, others

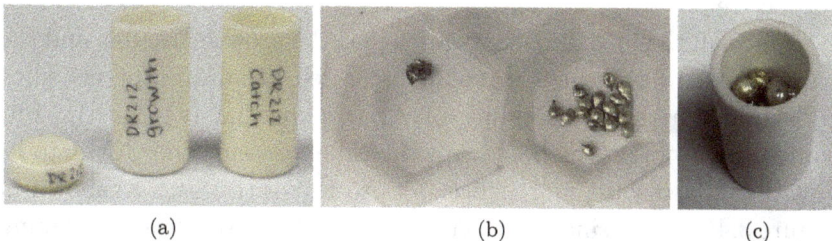

Figure 3.3.   (a) Canfield Crucible Set, CCS, labeled; (b) pieces of Pt and Sn after weighing; (c) reagents inside the crucible.

as powders with different granularity, it can sometimes be difficult to estimate this volume precisely. Basically, you want to (i) avoid having any reagent touch the bottom of the frit, but also (ii) not have only 10% of the volume of the crucible used, since there will not be much liquid volume to grow the crystals out of then. If you do not have experience with the density of the element that you plan use as the majority constituent (i.e. Sn for this growth), we recommend that completely fill one crucible with many pieces as possible, without pressing, to have a notion of how much mass you may want to use.

We can now weigh and label all the three pieces of CCS that will be used, as shown in Figure 3.3(a). The growth side is where we will place the element and grow the crystals; the catch side is where the decanted liquid will go after passing through the frit, during the decanting process at the end of the growth, much like water through a kitchen strainer [18]. Knowing these weights can allow for determination of how much material was decanted and how much remained on the growth side. Next weigh out the majority element, Sn in this case. With that mass known, you can calculate how much Pt is needed and weigh it out as well (Figure 3.3(b)). With this done the elements can be put inside the growth crucible; put some of Sn first, then Pt and then the rest of Sn. This ensures that the liquid Sn covers the Pt pieces increasing the area of contact (Figure 3.3(c)). Therefore, for this growth, in a 2 ml Al_2O_3 crucible at initial composition of $Pt_{04}Sn_{96}$, we used 3.0921 g of Sn and 0.2117 g of Pt, the mass of growth crucible was $m_G = 5.0574$ g and the mass of the catch crucible and frit were $m_C = 5.0816$ g and $m_F = 1.9784$ g, respectively.

Once all the initial reagents are inside the growth crucible and the CCS is assembled, the set should be sealed in an amorphous silica tube under a partial pressure of Ar (Figures 3.4 (a)–3.4(d)). Put some silica wool on the bottom of the tube, pressing the CCS into it, and then place some more silica wool to top of the CCS as shown in Figure 3.4(a). The silica wool at the bottom helps to prevent breaking of the tube all due to the difference of thermal contraction/expansion between the crucible and the tube. The silica wool in the upper part of the tube serves as a cushion for the crucible set when it goes

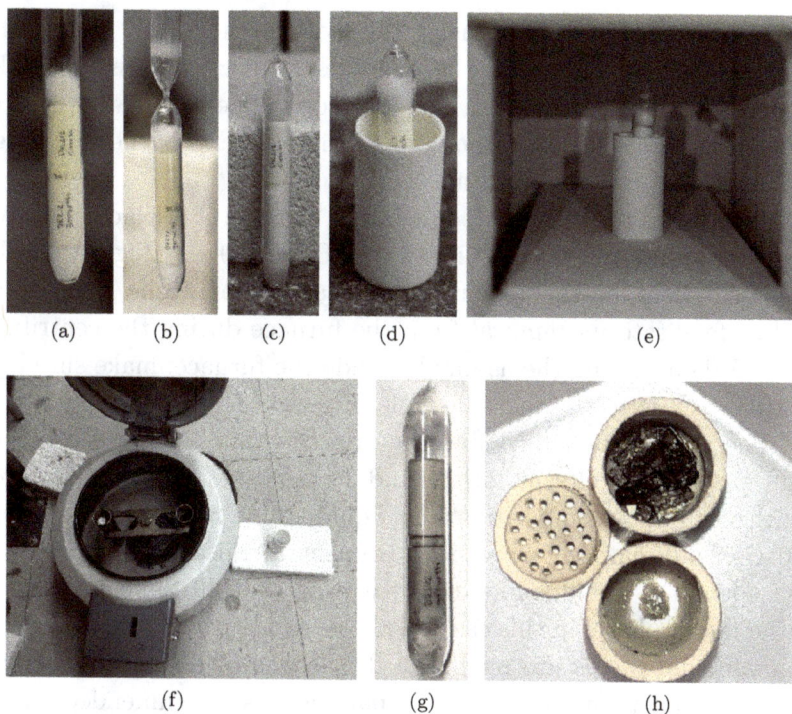

Figure 3.4. (a) Assembly of CCS in the silica tube; (b) necked ampoule before sealing; (c) sealed ampoule; (d) ampoule in the 50 ml crucible; (e) position of the assembly inside the furnace; (f) centrifuge with ampoule in left side; (g) ampoule after decanting process; (h) growth crucible (top) with $PtSn_4$ single crystals and catch crucible (bottom) with remaining flux solidified.

through the centrifugation process. To progress from Figure 3.4(a) to Figure 3.4(b), a blow torch is used to heat the silica tube until it collapses close to the upper silica wool plug but taking care not to completely close the tube, i.e., leaving a small neck for the passage of air. During this "necking" process, care must be taken to avoid having any of the reagents fall out of the growth crucible. Ideally, if the CCS is well embedded in the silica wool during this process such mishaps can be avoided. The next step is to place the necked tube onto a vacuum system for sealing. First pump out the atmosphere (50 Mtorr is a fine pressure), followed by a back filling with Ar. This pump/refill with Ar process should be performed multiple times (three in our group) before using the blow torch to fully sealing the ampoule. While sealing off the ampoule, you want to try to make sure that the sealed tip is not too sharp or fragile, Figure 3.4(c); and the length of the sealed ampoule fits inside the cup in the centrifuge that will be used for decanting.

In order to place the ampoule in the furnace, it is placed inside a larger crucible, in our case it was Al_2O_3 of 50 ml, Figure 3.4(d). This will help in the placing of the ampoule in the furnace as well as facilitate its removal from the furnace during the centrifuge stage. When placing the assembly inside the furnace, make sure it is relatively close to the thermocouple, Figure 3.4(e).

The next step is to define the growth ramp. As we can see in Figure 3.1, Sn melts at 231°C and Pt melts at 1769°C, however for the $Pt_{04}Sn_{96}$ composition, the liquidus line is at approximately 520°C. As discussed before, it is prudent to raise the temperature well above the liquidus line, thus ensuring that all the elements are in the liquid phase, in this case we chose to heat to 600°C. To ensure that all solid pieces are melted, a dwell time of several (2–5 hours depending on temperature and impatience) is recommended. After the dwell, a slow cooling of the melt is used to reduce the number of nucleation sites and, ideally lead to a small number of well-formed crystals. In this case, we cooled over 60 hours.

The last decision associated with the cooling process is to decide at what temperature to remove the remaining liquid from the crystals. In the range of composition between $Pt_{05}Sn_{95}$ to Pt_0Sn_{100}, the minimal temperature that still has liquid is 231°C. Whereas we could cool down to a few tens of degrees above this temperature, for this example the temperature at which the ampoule was removed from the furnace was 350°C [9].

Now for the fun and tricky part: the centrifugation step must be performed as rapidly, and yet safely, as possible. If you are a beginner, decanting at temperatures as low as 300°C may seem daunting. As you become more experience, decanting at temperatures as high as 1150°C will become possible. We recommend, at the beginning, that this process be done by two people: one person is responsible for opening and closing the furnace door while the other removes the ampoule and places it in the centrifuge. If you are a beginner, you should practice beforehand with all the safety apparatus and the furnace at room temperature, using "dummy tubes" until you feel confident with the process. For the safety of all involved in the process, we recommend to wear long pants, closed shoes, lab coat, gloves for high temperatures and a face-shield. If the temperature of the removal the ampoule was above $700°C$ use a partially reflective, i.e. Au-coated, face-shield.

The centrifuge we used, Figure 3.4(f), had a rotor built with two metal cups: one for the ampoule and the other for a counterbalance mass (often small coins from around the world). On the ampoule side we first place a considerable amount of fibrous furnace insulation at the bottom of the cup to act as a cushion or shock absorber for the sealed tube. Now, all is in readiness. It is time to decant. The furnace door is opened, the 50 ml crucible is grasped with tongs/forceps and removed from the furnace and brought to the centrifuge. The hot ampoule is transferred, catch side down, into the rotor cup, the lid of the centrifuge is closed and it is turned on. At this point, it will become apparent that it is important to also have a refractory brick next to the centrifuge. It will be helpful to place the hot, 50 ml crucible on it during the decanting process. Over the course of 10–15 s, it accelerates toward a target rotation of several 1000 rpm and then it is shut off. Once it has come to a stop the lid can be opened. Remove the ampoule from the centrifuge and wait until its temperature is close to room temperature, Figure 3.4(g).

The growth can now be opened and you can finally see if crystals have grown (Figure 3.4(h)). Once opened, be sure to weigh the three parts of the CCS and the crystals obtained. This will allow you have a good estimate of what your yield was, and even put points on the binary phase diagram [18]. With crystals in hand, measurements of field and temperature-dependent magnetization, resistivity, specific heat, etc. are all possible.

3.3.2. *BaBi₃: An air-sensitive growth*

All the process described for the growth of $PtSn_4$ need to be follow for the growth of $BaBi_3$, but since the elemental Ba and as well as the resultant $BaBi_3$ are both sensitive to air, the steps need to have

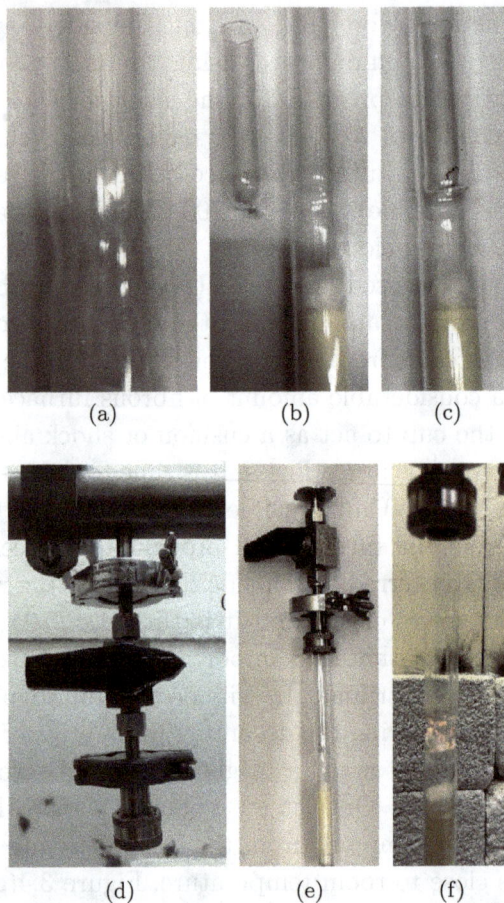

Figure 3.5. (a) Silica tube with slight constriction; (b) smaller plug and main tube of silica; (c) smaller tube inside the main one resting on constriction; (d) closeup of the valve used in the purge system; (e) setup removed from the glove box; (f) main tube being sealed, note glowing heated area.

modifications to allow for the manipulation of them in a glove box. Both elements are weighed in the glove box, and placed in a CCS. The difficulty now is how to get the CCS from the inert atmosphere of the glove box into the sealed ampoule without being exposed to air. The answer it shown in Figure 3.5. We use a multitube sealing system combined with a pumping flange/valve adapter for our pumping system. The most difficult step in this growth is sealing the silica ampoule with this whole assemblage on it. To facilitate the closure of the silica tube, we make modifications to the process.

We heat the silica tube that will form the ampoule so as to create a small constriction near what will be the top of the ampoule. This constriction (Figure 3.5(a)) will allow the CCS to pass but be sufficient to hold another, smaller silica tube in place. We prepare this second smaller silica tube, Figure 3.5(b), so that it gets stopped at the constriction Figure 3.5(c). Having this second silica tube above the packed CCS (Figures 3.5(e)–(f)) allows for sealing without the need for necking the tube, something that would be difficult with a flange/valve assemblage in place. Note that this preparation of the tubes has to be performed before placing them in the glove box.

Then, in the glove box, the CCS is packed into the larger silica tube with silica wool below and above it. The smaller tube is then lowered into place and the valve/flange assemblage is places on top. This then seals off the elements from air. The whole assemblage shown in Figure 3.5(e) is then cycled out of the glove box and attached to the pumping system. Once vacuum is established, the valve can be opened and the ampoule can be pumped and refilled with Ar as described above.

We heat the tube in the region where the two tubes are, so that when collapsing, the two tubes become soft and seal the ampoule, as shown in Figure 3.5(f). Since there is a very small distance between the two tubes, the sealing process is simplified. Once sealed, the ampoule can be removed from the rest of the tube. This process is what we use to seal ampoules containing elements sensitive to air and also toxic elements as in the case of growths containing arsenic. A simpler setup using a balloon, rather than the valve, is described in the appendix of Reference [23]: general tips and tricks for optimizing single crystal flux growths in [23].

The remainder of the growth process will be identical to that described in the previous example for $PtSn_4$. Once the ampoule is decanted and cooled, though, it has to be brought back into the glove box for opening.

3.4. Go Forth and Multiply (The Number of Known Compounds)

At this point, you hopefully have some feeling for what is entailed in the growth of binary compounds. Solution growth is a powerful and versatile technique and can allow you to create materials suitable for almost any measurement technique. If you indeed plan to try growing intermetallic compounds, in addition to the list of references provided in this text [9–24], we strongly urge you to collaborate with your institution's health and safety department or group in setting up a growth protocol that will keep you and others safe. Often a visit to a lab that has some or all of these facilities up and running can save a lot of time and prevent you from having to recreate the wheel. In our experience, researchers engaged in new materials growth and discovery can be like cooks, fishermen or joggers [25–27] and often are eager to share stories of innovations, triumphs or even impossible dreams [28]. Ultimately, we hope that you use this chapter as a jumping off point and share in our enjoyment of exploration and discovery.

Acknowledgments

Both RAR and PCC thank all of the students and colleagues who have worked with them through the decades. The discovery, growth, characterization and understanding of novel materials is group activity and part of its joy is associated with the diversity and idiosyncrasies of the group.

This work was supported by the U.S. Department of Energy, Office of Basic Energy Science, Division of Materials Sciences and Engineering. The research was performed at the Ames Laboratory. Ames Laboratory is operated for the U.S. Department of Energy by Iowa State University under Contract No. DE-AC02-07CH11358. RAR was funded by the Gordon and Betty Moore Foundation's EPiQS Initiative through Grant GBMF4411.

Bibliography

[1] K. L. Wang, M. Lang, and X. Kou, in *Handbook of Spintronics*, Y. Xu, D. D. Awschalom and J. Nitta (eds.), (Springer, 2016), pp. 431–462.

[2] M. König, S. Wiedmann, C. Brüne, A. Roth, H. Buhmann, L. W. Molenkamp, X.-L. Qi, and S.-C. Zhang, *Science*, **318**, 766 (2007).

[3] H. Zhang, C.-X. Liu, X.-L. Qi, X. Dai, Z. Fang, and S.-C. Zhang, *Nature Phys.*, **5**, 438 (2009).

[4] Y. Xia, D. Qian, D. Hsieh, L. Wray, A. Pal, H. Lin, A. Bansil, D. Grauer, Y. S. Hor, R. J. Cava, and M. Z. Hasan, *Nature Phys.*, **5**, 398 (2009).

[5] S. Kim, M. Ye, K. Kuroda, Y. Yamada, E. E. Krasovskii, E. V. Chulkov, K. Miyamoto, M. Nakatake, T. Okuda, Y. Ueda, K. Shimada, H. Namatame, M. Taniguchi, and A. Kimura, *Phys. Rev. Lett.*, **107**, 056803 (2011).

[6] W. Liu, X. Peng, C. Tang, L. Sun, K. Zhang, and J. Zhong, *Phys. Rev. B*, **84**, 245105 (2011).

[7] B. Yan, and C. Felser, *Annu. Rev. Condens. Matter Phys.*, **8**, 337 (2017).

[8] N. P. Armitage, E. J. Mele, and A. Vishwanath, *Rev. Mod. Phys.*, **90**, 015001 (2018).

[9] E. Mun, H. Ko, G. J. Miller, G. D. Samolyuk, S. L. Bud'ko, and P. C. Canfield, *Phys. Rev. B*, **85**, 035135 (2012).

[10] Y. Wu, L.-L. Wang, E. Mun, D. D. Johnson, D. Mou, L. Huang, Y. Lee, S. L. Bud'ko, P. C. Canfield, and A. Kaminski, *Nature Phys.*, **12**, 667 (2016).

[11] R. Jha, M. A. Avila, and R. A. Ribeiro, *Supercond. Sci. Technol.*, **30**, 025015 (2017).

[12] L. Xiang, R. A. Ribeiro, U. S. Kaluarachchi, E. Gati, M. C. Nguyen, C.-Z. Wang, K.-M. Ho, S. L. Bud'ko, and P. C. Canfield, *Phys. Rev. B*, **98**, 214509 (2018).

[13] N. Ni, M. E. Tillman, J.-Q. Yan, A. Kracher, S. T. Hannahs, S. L. Bud'ko, and P. C. Canfield, *Phys. Rev. B*, **78**, 214515 (2008).

[14] W. R. Meier, T. Kong, S. L. Bud'ko, and P. C. Canfield, *Phys. Rev. Mater.*, **1**, 013401 (2017).

[15] H. Okamoto, *Desk Handbook: Phase Diagrams for Binary Alloys*, (ASM International, 2000).

[16] P. C. Canfield, M. L. Caudle, C.-S. Ho, A. Kreyssig, S. Nandi, M. G. Kim, X. Lin, A. Kracher, K. W. Dennis, R. W. McCallum, and A. I. Goldman, *Phys. Rev. B*, **81**, 020201(R) (2010).

[17] T. Yamada, H. Takakura, T. Kong, P. Das, W. T. Jayasekara, A. Kreyssig, G. Beutier, P. C. Canfield, M. de Boissieu, and A. I. Goldman, *Phys. Rev. B*, **94**, 060103(R) (2016).

[18] P. C. Canfield, T. Kong, U. S. Kaluarachchi, and N. H. Jo, *Philos. Mag.*, **96**, 84 (2016).

[19] N. Haldolaarachchige, S. K. Kushwaha, Q. Gibson, and R. J. Cava, *Supercond. Sci. Technol.*, **27**, 105001 (2014).
[20] P. C. Canfield (ed.), *Philos. Mag.*, **92**, 2397–2711 (2012).
[21] I. R. Fisher, M. C. Shapiro, and J. G. Analytis, *Philos. Mag.*, **92**, 2401 (2012).
[22] P. C. Canfield, and I. R. Fisher, *J.Crystal Growth*, **225**, 155 (2001).
[23] R. A. Ribeiro, and M. A. Avila, *Philos. Mag.*, **92**, 2492 (2012).
[24] P. Cedomir, P. C. Canfield, and J. Y. Mellen, *Philos. Mag.*, **92**, 2448 (2012).
[25] P. C. Canfield, *Nature Phys.*, **5**, 529 (2009).
[26] P. C. Canfield, *Nature Phys.*, **4**, 167 (2008).
[27] P. C. Canfield, *APL Mater.*, **3**, 041001 (2015).
[28] P. C. Canfield, *Rep. Prog. Phys.*, **83**, 016501 (2020).

Chapter 4

Vapor Transport Growth of van der Waals Magnets

A. V. Haglund*,‡, J. Q. Yan*,†,§, and D. G. Mandrus*,†,∥

*Department of Materials Science and Engineering, University of Tennessee, Knoxville, TN, 37996, USA
†Materials Science and Technology Division, Oak Ridge National Laboratory, Oak Ridge, TN, 37831, USA
‡ahaglund@vols.utk.edu
§yanj@ornl.gov
∥dmandrus@utk.edu

4.1. Introduction to van der Waals Magnets

A new major area of materials research opened up after the revolutionary isolation of atomically thin graphene sheets in 2004 [1]: layered van der Waals materials. The first of many layered materials to be recognized with this ability, graphene is composed of layers of carbon that are a single atom thick, with the sheets bonded together only by weak forces. This allows the layers to be separated easily with something as simple as Scotch tape, paving the way for new technology that could be based on single layers of material. Also known as 2D materials for their two-dimensional (2D) structure, many more layered materials have been predicted, created, or rediscovered in the literature, with layers consisting of either single atoms or a compound structure of multiple elements. The family of layered materials ranges widely in properties, including: wide-band gap insulators [2], semiconductors [3], metals [4], and magnets [5, 6].

Layered materials with magnetic properties have been termed van der Waals magnets, and come in many varying compositions, structures, and properties based on transition metal and rare-earth ions, and are typically composed of the metal ions sandwiched between chalcogens or halogens. Some examples are transition metal dichalcogenides such as room-temperature ferromagnetic $CrTe_2$ [7], semiconducting transition metal trichalcogenides such as antiferromagnetic $CrTe_3$ [3] or ferromagnetic $CrSiTe_3$ [5], transition metal thiophosphates and selenophosphates such as antiferromagnetic MPX_3 (M = transition metal, X = S or Se) [8], transition metal halides such as ferromagnetic CrI_3 [9], and rare-earth halides such as high magnetoresistance, high-Tc ferromagnetic GdI_2 [10]. A more extensive list has been collected in the two review articles [11, 12], covering all known categories of magnetic layered materials.

These 2D magnetic materials are interesting because of their potential for both new physics and technologies. Some significant phenomena have been discovered in these materials, including the proximate Kitaev spin-liquid α-$RuCl_3$ [13] and room temperature ferromagnetism in monolayer $MnSe_2$ [14]. Layered material behavior can also be altered by reducing the number of layers in the material down to few- or single-layers, such as the change in Curie temperature for $CrGeTe_3$ [15] and CrI_3 [16]. Some applications currently being investigated for these materials include single layers employed in spintronic devices [17] or van der Waals heterostructures. The latter research involves stacking monolayers of the same or different materials, with the idea being to design a new material with tailored properties [18]. More potential areas for application are continuously being discovered [12]. As the limit of Moore's law is rapidly approaching with conventional transistors, these new materials can be the necessary fuel to push technology forward.

Since the field of atomically thin layered materials is still relatively new, basic research of the materials and their applications is still in progress. Although some materials can already be deposited in single layers by molecular beam epitaxy (MBE) like $MnSe_2$ [14], or by chemical vapor deposition (CVD) for $MoTe_2$ and WTe_2 [19, 20] among others [21], these methods have not yet been formulated for the majority of 2D materials. To transition to material applications beyond basic science, wide area large-scale production will be necessary [11]. Until these synthesis methods are discovered for all 2D materials, research can continue through exfoliation of layers from single crystals, since many bulk-crystal growth methods are already available in the literature. In many cases, exfoliation of bulk crystals may lend itself better to pioneering research. Although some work has been done with forming heterostructures through deposition methods, summarized well by Zhang *et al.* [21], there are many more material combinations that have not yet been explored. In the meantime, stacking layers of different compositions to form heterostructures can currently be done with exfoliated layers to help determine which material combinations would be useful to pursue later. Regardless of the present limits on wide-area synthesis, a variety of single-crystal growth methods currently exist, allowing synthesis of any material in the bulk form that is known to grow.

Multiple methods exist for growing bulk single crystals of magnetic layered materials, including vapor transport [22], flux growth [23–25], and the Bridgman technique [26]. While all these methods are employed to grow 2D materials, vapor transport can be used to synthesize the largest variety due to a greater number of the compounds containing chalcogens and halogens and will therefore be the growth technique covered in this chapter.

4.2. Vapor Transport Growth

4.2.1. *Introduction to vapor transport technique*

Vapor transport is one of the main methods used to purify starting materials and grow single crystals of layer-structured compounds. During crystal growth, the starting solid phase, normally in powder form, is vaporized, transported, and deposited elsewhere in the form of single crystals, through the influence of an applied temperature gradient. This growth can occur in the presence of a gaseous phase called the transport agent. As illustrated in Figure 4.1, the transport agent, composed of a volatile compound, reacts with the solid phase material, $A(s, \text{poly})$, in the charge end of the tube, producing a gas, $AX(g)$, with high vapor pressure which is in equilibrium with the solid. As the gas fills the tube and travels through the temperature

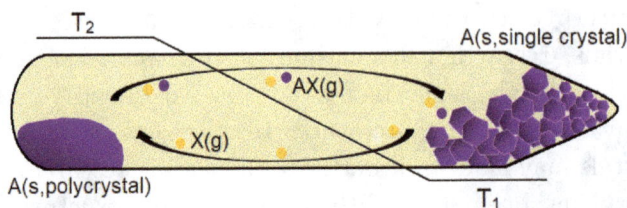

Figure 4.1.   Single crystal growth from a polycrystalline solid phase (A) by vapor transport inside a fused quartz ampoule with a transport agent (X) under a temperature gradient ($T_2 > T_1$). Figure inspired by [22].

gradient, it falls out of equilibrium at T_1, reversing the reaction and depositing the solid, which nucleates and grows single crystals. Once the transport agent has deposited its charge, it disperses back to begin the process again. Typically, the transport direction is from a region of high temperature (T_2) to low temperature (T_1), indicating an endothermic reaction between the charge (A_s) and transport agent (X_g), but sometimes the opposite occurs as in the case of exothermic reactions [22].

This process with the aid of a transport agent is termed chemical vapor transport (CVT). In some cases, transport and crystal growth can happen without the addition of other chemicals as a transport agent. Called physical vapor transport (PVT), this method occurs through a few different routes such as sublimation, decomposition sublimation, or auto transport. PVT works well for transition metal halides and chalcogenides [27], as these compounds tend to be volatile or decompose more easily into volatile gases, producing the necessary vapor pressure for self-transport.

4.2.2. *Starting materials*

The starting materials for growth can be either the constituent elements in stoichiometric amounts or presynthesized compound, typically in powder form to increase the reaction surface area and transport rate. Generally, 1–2 grams are enough to produce a significant number of crystals in a single transport reaction [28]. Limiting the initial charge to this amount is also helpful for avoiding vapor overpressure and explosion of the tube. Excessive vapor pressure is mostly a concern with sulfur-containing compounds, but selenium and some transport agents can also cause this problem. If necessary, before starting a growth the vapor pressure can be estimated by determining the amount and type of gaseous substances evolved during the transport reaction and calculating the pressure from $PV = nRT$ at the reaction temperature, to ensure a reasonable pressure. For other situations, such as when presynthesizing charge materials, heating slowly to the compound formation temperature can help powdered elements to react and form compounds while minimizing the amount of volatilized material at one time. Larger charge pieces of the elements instead of powder can also slow evaporation of materials, if tube overpressure is a persistent problem.

4.2.3. *Transport agent*

For the CVT process, the external transport agent consists of an easily-vaporized compound, with a number of options to choose from depending on the material grown, for example, I_2, Br_2, Cl_2, S, Se, $TeCl_4$, etc. [28]. The amount of transport agent for the reaction is calculated for the desired concentration and tube size, and is then loaded into the tube along with the charge material. For magnetic layered materials grown by CVT such as $MnPS_3$ [29] or Fe_3GeTe_2 [30], vapor transport is often performed with iodine. Transport agent amounts vary greatly depending on the temperature and the ease with which the material transports, but iodine concentrations are typically between 2–12 mg/cm^3 [31, 32].

4.2.4. *Growth ampoule preparation*

Before sealing the quartz tube with the materials inside, there are some precautions that should be taken. If there are concerns about impurities on the inside tube surface, the tube can be rinsed with a solution of 1:1 nitric acid/water, which also helps reduce surface roughness. This is important because surface roughness and impurities act as nucleation sites for crystal growth, where an excess can limit the final crystal size if too many crystals nucleate and grow. Once the tube is cleaned, the charge and transport agent, if required for the reaction, are loaded into the bottom, while being careful not to deposit material along the tube surface. A rolled-up piece of paper can be used to line the inside of the tube while it is being filled, and then removed once done. Particles on the inner surface of the tube may make tube-sealing more difficult, as they can cause the tube to fracture while sealing by reacting with the quartz as it is heated. After attaching the loaded tube to a vacuum system, it is carefully flushed multiple times with inert gas and then brought to high vacuum to remove oxygen and moisture. Either an ice-bath, wet paper towel, or some other form of cooling is frequently used to surround the charge end of the tube during tube-sealing, to reduce loss of easily volatilized material to the vacuum. A can of "Dust-Off" can also be used to cool the tube before sealing; holding the can upside down allows the liquid to decompress instead of gas and will rapidly cool the tube. Once the charge is protected from heat, the tube is then locally heated with an oxygen–hydrogen torch at a predetermined length from the charge, usually 10–20 cm, and then sealed to create a vacuum-protected growth ampoule.

4.2.5. *Furnace choice*

Before designing a crystal growth process, furnace type is considered since available temperatures and gradients will be dependent on furnace capabilities. Depending on the precision of the growth temperature requirements and temperature gradients, one of the three kinds of furnaces is typically used: either a multi- or single-zone horizontal tube furnace, or a box furnace. Multi-zone tube furnaces allow multiple areas of the tube to be separately temperature controlled, which is important if the formation temperature range for a certain crystal structure or compound is limited. This allows for precise control of the temperature gradient as well. Single-zone furnaces are less precise, since the temperature at the center of the furnace is set but the ends are open, allowing for a natural temperature gradient. This format has some versatility, as the gradient and growth temperature are controlled by adjusting the tube length instead. Box furnaces are used when only a minimal temperature gradient is needed, by taking advantage of the slight temperature gradients inside the furnace.

4.2.6. *Growth parameters*

Many factors are involved in designing a successful crystal growth environment, including the growth temperature, charge temperature, temperature gradient, transport agent, and time. To design the growth of a new compound, often vapor transport reactions of similar compounds are reviewed to figure out the general growth parameters needed. If vapor transport reactions of similar starting materials do not exist, possible reactions for the compound formation are determined by looking at their decomposition equations or their reactions with various transport materials [28]. Once a general idea of temperature and starting materials is known, it is helpful to check reaction tables of all elements or compounds involved to make sure no other competing compounds will form solids at the same temperatures [33]. However, even well-known growth reactions usually require some fine-tuning to optimize the growth rate and crystal size. If the transport is too slow and temperature is the issue, this can be resolved by either increasing the average temperature of the growth, increasing the temperature gradient across the tube, or raising the growth-end temperature [22, 31].

While lengthening the tube can be used to increase the temperature difference and therefore the temperature gradient in single-zone type furnaces, changing the tube diameter can also have an impact on the transport rate in any furnace type, in addition to effects on the crystal size and quality [31, 32]. Inner tube diameters typically range from 10 to 20 mm, but larger diameter tubes are often chosen for their greater transport rate due to higher convection effects. Crystals also tend to grow larger and less intertwined with one another in wider tubes; this allows for easier separation and a greater number of useable crystals from the initial charge material.

For vapor transport reactions requiring a transport agent, the growth rate can also be accelerated by increasing the amount of transport agent. Reported in a study by Whitehouse and Balchin, the transport rate as well as crystal size and quality increased with iodine transport concentrations up to 10 mg/cm^3 [31]. Generally, 2 mg/cm^3 seems to be a preferred starting point, though optimal concentrations vary from one material to the next.

The growth time required to reach the desired crystal size varies, depending on the material and the factors previously mentioned, with durations ranging from just a few hours to a few months. If too many crystals grow at once and limit the final crystal size, even after tube cleaning, reversing the temperature direction of the tube for the first day of growth can help eliminate some of the excess nucleation sites by transferring traces of charge powder on the sides of the tube back to the charge end [22].

Table 4.1 shows some detailed examples of vapor transport growth for various materials, including CVT, decomposition sublimation, and auto-transport.

Figure 4.2 contains pictures of some crystals grown in our lab. MPX_3 crystals were prepared based on growth details from [29]. Initial compound was synthesized as powder by measuring out each element in a stoichiometric ratio, grinding together to form a homogeneous mixture under inert atmosphere, and then pressing into a pellet. Each pellet was sealed under vacuum in a quartz tube and annealed for one week at 730°C to form $MnPS_3$, $MnPSe_3$, $FePS_3$, $FePSe_3$, and at 700°C for $NiPS_3$, and then the CVT growth details shown in Table 4.1 were followed. $CrPS_4$ crystals were also prepared in our lab with good results; the details are shown in the table and are loosely based on directions given in [35]. Most compounds in the table were synthesized in single-zone tube furnaces, although multizone furnaces are used as well. Crystals of α-$RuCl_3$ have also been grown by vapor transport with $TeCl_4$ as the transport agent [36], or by sublimation at high temperatures [37].

4.3. Flux Growth vs. Vapor Transport

Due to the large vapor pressure of halides, chalcogenides, selenophosphates, and thiophosphates, vapor transport has been widely used to grow these layered exfoliable materials. As described in Part 2,

Table 4.1.   Examples of vapor transport growths.

Compound	Starting materials	Transport method	Tube length/diameter	Temperature gradient	Growth time	References
Fe$_3$GeTe$_2$	Fe:Ge:Te, 3:1:2	I$_2$, 2 mg/cc	—	750°C–700°C	1 week	[30]
MnPS$_3$	MnPS$_3$powder	I$_2$, 10–15 mg/cc	13 cm/1.9 cm	700°C–570°C	<1 week	[29]
MnPSe$_3$	MnPSe$_3$ powder	I$_2$, 10–15 mg/cc	13 cm/1.9 cm	650°C–525°C	<1 week	[29]
FePS$_3$	FePS$_3$ powder	I$_2$, 10–15 mg/cc	10 cm/1.9 cm	700°C–600°C	<1 week	[29]
FePSe$_3$	FePSe$_3$ powder	I$_2$, 10–15 mg/cc	10 cm/1.9 cm	670°C–570°C	1–2 weeks	[29]
NiPS$_3$	NiPS$_3$ powder	I$_2$, 10–15 mg/cc	10 cm/1.9 cm	700°C–600°C	1–2 weeks	[29]
CrI$_3$	Cr:I, 1:3 ratio	Decomposition Sublimation	15 cm/2.2 cm	650°C–500°C	1–2 weeks	[34]
CrPS$_4$	Cr:P:S, 1:1:4 powder & chunks	Auto-Transport	13 cm/1.9 cm	700°C–570°C	1–2 weeks	[35]

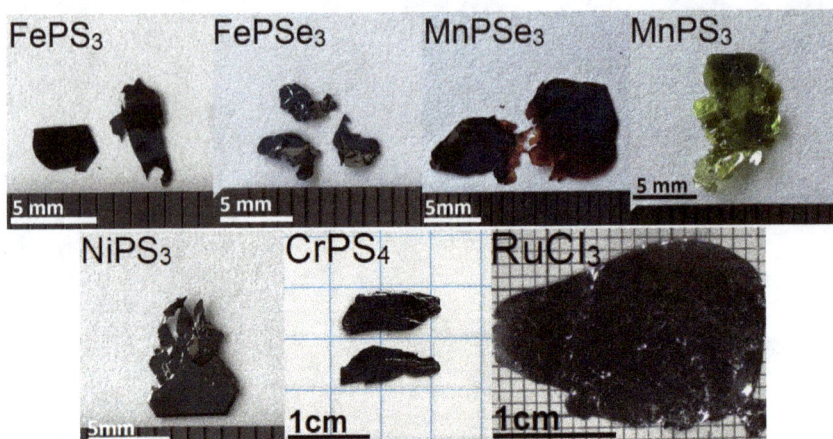

Figure 4.2.   Single crystals of layered materials grown by vapor transport.

PVT involves the sublimation and recrystallization of starting materials, making good use of their large vapor pressure at elevated temperatures. PVT is the ideal technique for the growth of various transition metal halides. Chemical vapor transport, where a small amount of transport agent is normally needed, works better for the growth of chalcogenides, selenophosphates, and thiophosphates.

Single crystals of transition metal tellurides can be grown using vapor transport technique, flux growth, or Bridgman technique. For most tellurides, both vapor transport technique and flux growth provide crystals of the same stoichiometry and similar quality. However, in some special cases, the flux growth in a tellurium-rich environment can affect the stoichiometry of as-grown crystals. One good example is $ZrTe_5$: crystals grown by vapor transport technique tend to be Te deficient, while flux growth out of Te melt normally leads to nearly stoichiometric crystals [38]. Table 4.2 provides some examples of crystal growth for Te-based van der Waals magnets.

One more situation when flux growth is preferred is when large telluride crystals are needed for some special measurements such as neutron scattering. The exfoliable tellurides tend to crystallize into thin plates in vapor transport growths, in which the cross-section

Table 4.2. Tellurium-based flux growth details.

Material	Molar ratio	Heating rate	Solution time/ temperature	Cooling rate	Final cool	References
CrSiTe$_3$	Cr:Si:Te 1:2:6	1150°C @ 200°C/hr	16 hrs @ 1150°C	3°C/hr	Centrifuge @700°C	[39]
CrGeTe$_3$	Cr:Ge:Te 1:2:9	800°C @ 200°C/hr	10 days @ 800°C	5.55°C/hr	Centrifuge @550°C	[23]
Fe$_{3-x}$GeTe$_2$	Fe:Ge:Te 2:1:4	950°C @ 200°C/hr	12 hrs @ 950°C	1°C/hr or 3°C/hr	Centrifuge @675°C	[40]
CrTe$_3$	Cr:Te 1:99	—	16 hrs @ 1050°C	1.5°C/hr	Centrifuge @455°C	[3]
	3:97	—	24 hrs @ 600°C			

Figure 4.3. CrSiTe$_3$ single crystals grown out of excess Si & Te.

of the transported crystals is generally small and limited by the diameter of the quartz ampoule. In contrast, flux growth can sometimes provide thick telluride plates. As shown in Figure 4.3, large CrSiTe$_3$ single crystals around 2g per piece can be grown out of Te-rich melt, which make inelastic neutron scattering studies possible [5].

For an overview of the variety available in magnetic layered materials, some example compounds with a range of compositions

Table 4.3.   Some examples of van der Waals magnets and properties.

Compound	Magnetic ordering	T_C or T_N	Calculated cleavage energy (J/m^2)	Insulating/ metallic
MnSe$_2$ — monolayer	FM [14]	>300 K [14]	—	I [14]
1T-VSe$_2$ — monolayer	FM [41]	>330 K [41]	0.36[42]	M [41]
*Fe$_3$GeTe$_2$	FM [30]	150 K [40] 220 K [30]	—	M [30]
Fe$_5$GeTe$_2$	FM [43]	310 K [43]	—	M [43]
CrSiTe$_3$	FM [39]	33 K [39]	0.35–0.44 [44, 45]	I [39]
CrGeTe$_3$	FM [46]	61 K [46]	0.38 [44]	I [47]
CrTe$_3$	AFM [3]	55 K [3]	0.5 [3]	I [3]
CrPS$_4$	AFM [35]	36 K [35]	—	I [35]
MnPS$_3$	AFM [48]	78–82 K [6, 8, 48]	0.26 [49]	I [50]
MnPSe$_3$	AFM [48]	74 K [48]	0.24 [51]	I [52]
FePS$_3$	AFM [8]	111–123 K [6, 8, 53, 54]	0.26 [49]	I [50]
FePSe$_3$	AFM [55]	106–123 K [8, 55, 56]	0.37 [49]	I [57]
NiPS$_3$	AFM [54]	151–155 K [54, 58]	0.22 [49]	I [50]
NiPSe$_3$	AFM [59]	206 K [59]	—	—
CrI$_3$	FM [34]	61 K [34]	0.29–0.30 [34, 60]	I [34]
α-RuCl$_3$	AFM [13]	7 K [13]	—	I [13]
GdI$_2$	FM [61]	280–295 K [10, 61]	—	M [61]
YbCl$_3$	AFM [62]	0.60 K [62]	—	I [62]

* Transition temperature can be finely tuned by varying the Fe content [40].

and structures are shown in Table 4.3 along with their related properties.

For the full range of van der Waals magnets, the review articles [11, 12] are quite thorough. Known transition metal dihalide and trihalide compounds have also been compiled in a recent review article by McGuire *et al.* [9] Several of these halide compounds have additionally been mentioned in another review by Lieth [27], with some useful references to their growth methods.

4.4. Summary

As seen in this chapter, vapor transport is a versatile growth method that can be used to produce many different layered magnetic materials of varying compositions and properties, from chalcogenides to halides and wide bandgap materials to metals. With atomically thin van der Waals magnets being a relatively new field, many compounds are still missing not only large-scale production methods, but also in some cases complete characterization and understanding of the material properties and behavior. Important and necessary work can continue with both characterization and application research on single crystals grown through vapor transport, until other mass production techniques are developed.

Acknowledgments

A.V.H. and D.G.M. were supported by the Gordon and Betty Moore Foundation's EPiQS Initiative Grant No. GBMF4416. J.-Q.Y. acknowledges support from the U.S. Department of Energy (U.S.-DOE), Office of Science — Basic Energy Sciences (BES), Materials Sciences and Engineering Division.

Bibliography

[1] K. S. Novoselov *et al.*, *Science*, **306**, 666 (2004).
[2] K. Nakao, and N. Kurita, *J. Phys. Soc. Jpn.*, **58**, 610, (1989).
[3] M. A. McGuire *et al.*, *Phys. Rev. B Condens. Matter.*, **95**, 144421 (2017).
[4] H.-J. Deiseroth, K. Aleksandrov, C. Reiner, L. Kienle, and R. K. Kremer, *Eur. J. Inorg. Chem.*, **2006**, 1561 (2006).
[5] T. J. Williams *et al.*, *Phys. Rev. B Condens. Matter.*, **92**, 144404 (2015).
[6] P. A. Joy, and S. Vasudevan, *Phys. Rev. B.*, **46**, 5425 (1992).
[7] D. C. Freitas *et al.*, *J. Phys. Condens. Matter.*, **27**, 176002 (2015).
[8] G. Le Flem, R. Brec, G. Ouvard, A. Louisy, and P. Segransan, *J. Phys. Chem. Solids*, **43**, 455 (1982).
[9] M. A. McGuire, *Crystals*, **7**, 121 (2017).
[10] R. K. Kremer, and A. Simon, *Curr. Appl. Phys.*, **4**, 563 (2004).
[11] C. Gong, and X. Zhang, *Science*, **363**, (2019), doi:10.1126/science. aav4450.

[12] K. S. Burch, D. Mandrus, and J.-G. Park, *Nature*, **563**, 47 (2018).

[13] A. Banerjee *et al.*, *Science*, **356**, 1055 (2017).

[14] D. J. O'Hara *et al.*, *Nano Lett.*, **18**, 3125 (2018).

[15] C. Gong *et al.*, *Nature*, (2017), doi:10.1038/nature22060.

[16] B. Huang *et al.*, *Nature*, **546**, 270 (2017).

[17] T. Jungwirth, X. Marti, P. Wadley, and J. Wunderlich, *Nat. Nanotechnol.*, **11**, 231 (2016).

[18] A. K. Geim, and I. V. Grigorieva, *Nature*, **499**, 419 (2013).

[19] L. Zhou *et al.*, *J. Am. Chem. Soc.*, **137**, 11892 (2015).

[20] K. Chen *et al.*, *Adv. Mater.*, **29**, (2017), doi:10.1002/adma.201700704.

[21] K. Zhang, Y.-C. Lin, and J. A. Robinson, in: F. Iacopi, J. J. Boeckl, and C. Jagadish, (eds.), *Semiconductors and Semimetals* (Elsevier; 2016), pp. 189–219.

[22] P. Schmidt, M. Binnewies, R. Glaum, and M. Schmidt, in: S. Ferreira (ed.) *Advanced Topics on Crystal Growth* (InTech, 2013).

[23] L. D. Alegria, H. Ji, N. Yao, J. J. Clarke, R. J. Cava, and J. R. Petta, *Appl. Phys. Lett.*, **105**, 053512 (2014).

[24] D. H. Keum, *et al.*, *Nat. Phys.*, **11**, 482 (2015).

[25] J.-Q. Yan, B. C. Sales, M. A. Susner, and M. A. McGuire, *Phys. Rev. Mater.*, **1**, 023402 (2017).

[26] P. Meglino, and E. Kostiner, *J. Cryst. Growth*, **32**, 276 (1976).

[27] R. M. A. Lieth, *Preparation and Crystal Growth of Materials with Layered Structures*. E. Mooser, (ed.) (Springer Science+Business Media, 1977).

[28] M. Binnewies, M. Schmidt, and P. Schmidt, *Z. Anorg. Allg. Chem.*, **643**, 1295 (2017).

[29] W. Klingen, R. Ott, and H. Hahn, *Z. Anorg. Allg. Chem.*, **396**, 271 (1973).

[30] B. Chen *et al.*, *J. Phys. Soc. Jpn.*, **82**, 124711 (2013).

[31] C. R. Whitehouse, and A. A. Balchin, *J. Cryst. Growth*, **43**, 727 (1978).

[32] P. N. Dangel, and B. J. Wuensch, *J. Cryst. Growth*, **19**, 1 (1973).

[33] M. Binnewies, R. Glaum, M. Schmidt, and P. Schmidt, *Chemical Vapor Transport Reactions* (Walter de Gruyter, 2012).

[34] M. A. McGuire, H. Dixit, V. R. Cooper, and B. C. Sales, *Chem. Mater.*, **27**, 612 (2015).

[35] Q. L. Pei *et al.*, *J. Appl. Phys.*, **119**, 043902 (2016).

[36] M. Ziatdinov *et al.*, *Nat. Commun.*, **7**, 13774 (2016).

[37] S.-H. Do *et al.*, *Nat. Phys.*, **13**, 1079 (2017).

[38] P. Shahi *et al.*, *Phys. Rev. X.*, **8**, 021055 (2018).

[39] L. D. Casto *et al.*, *APL Mater.*, **3**, 041515 (2015).

[40] A. F. May, S. Calder, C. Cantoni, H. Cao, and M. A. McGuire, *Phys. Rev. B.*, **93**, (2016) doi:10.1103/PhysRevB.93.014411.

[41] M. Bonilla *et al. Nat. Nanotechnol.*, **13**, 289 (2018).
[42] T. Björkman, A. Gulans, A. V. Krasheninnikov, and R. M. Nieminen, *Phys. Rev. Lett.*, **108**, (2012), doi:10.1103/PhysRevLett.108.235502.
[43] A. F. May *et al.*, *ACS Nano.* (2019), doi:10.1021/acsnano.8b09660.
[44] X. Li, and J. Yang, *J. Mater. Chem.*, **2**, 7071 (2014).
[45] M.-W. Lin *et al.*, *J. Mater. Chem.*, **4**, 315 (2015).
[46] V. Carteaux, D. Brunet, G. Ouvrard, and G. Andre, *J. Phys. Condens. Matter.*, **7**, 69 (1995).
[47] H. Ji *et al.*, *J. Appl. Phys.*, **114**, 114907 (2013).
[48] P. Jeevanandam, and S. Vasudevan, *J. Phys. Condens. Matter.*, **11**, 3563 (1999).
[49] K.-Z. Du *et al.*, *ACS Nano.*, **10**, 1738 (2016).
[50] A. Kamata *et al.*, *J. Phys. Soc. Jpn.*, **66**, 401 (1997).
[51] X. Li, X. Wu, and J. Yang, *J. Am. Chem. Soc.*, **136**, 11065 (2014).
[52] V. Grasso, and L. Silipigni, *JOSA B* (1999), Available: https://www.osapublishing.org/abstract.cfm?uri=josab-16-1-132.
[53] J.-U. Lee *et al.*, *Nano Lett.*, **16**, 7433 (2016).
[54] R. R. Rao, and A. K. Raychaudhuri, *J. Phys. Chem. Solids*, **53**, 577 (1992).
[55] B. Taylor, J. Steger, A. Wold, and E. Kostiner, *Inorg. Chem.*, **13**, (1974), Available: http://pubs.acs.org/doi/pdf/10.1021/ic50141a034.
[56] P. Ferloni, and M. Scagliotti, *Thermochim. Acta.*, **139**, 197 (1989).
[57] R. Brec, D. M. Schleich, G. Ouvrard, A. Louisy, and J. Rouxel, *Inorg. Chem.*, **18**, 1814 (1979).
[58] A. R. Wildes *et al.*, *Phys. Rev. B Condens. Matter.*, **92**, 224408 (2015).
[59] R. Brec, and J. Rouxel, in: A. S. Strub and H. Ehringer (eds.) *New Ways to Save Energy.* (Commission of the European Communities, 1980), pp. 620–630.
[60] W.-B. Zhang, Q. Qu, P. Zhu, and C.-H. Lam, *J. Mater. Chem.*, **3**, 12457 (2015).
[61] C. Felser, K. Ahn, R. K. Kremer, R. Seshadri, and A. Simon, *J. Solid State Chem.*, **147**, 19 (1999).
[62] J. Xing *et al.*, *Phys. Rev. B*, **102**, 014427 (2020).

Chapter 5

Induction Furnace Heating for Growth of Intermetallic Quantum Materials

R. E. Baumbach

National High Magnetic Field Laboratory,
Florida State University, Tallahassee, Florida 32310, USA
Department of Physics, Florida State University,
Tallahassee, Florida 32306, USA
baumbach@magnet.fsu.edu

5.1. Introduction

Induction furnace heating has a broad utility for crystal growth in research settings and has been applied to produce a multitude of different materials [1–7]. This is because it has several advantages, including (1) that it enables rapid access to a broad range of temperatures, (2) provides active stirring of a molten charge, (3) offers the ability to precisely design hot zones and temperature gradients, (4) can be adapted to customized sample growth environments (inert gasses, reactive gasses, ultra-high vacuum), (5) is applicable for most standard bulk-crystal growth methods (e.g., direct reaction of elements, Czochralski crystal pulling, Bridgman crystal pulling, and molten flux growth), and (6) even allows for magnetic levitation of a molten charge (e.g., using a Hukin-style crucible) which completely eliminates the risk of chemical contamination from a crucible. Importantly, modern power supplies feature automatic frequency matching, which dynamically accounts for the work piece being part of a resonant circuit where the resonant frequency changes with time as the temperature and resistivity of the work piece evolve.

This optimizes the power transfer, and makes this method ideal for building custom crystal growth equipment that is optimized for specific techniques. In the following section, we briefly introduce the principle of induction furnace heating and describe its application using several methods in crystal growth apparatuses that have been constructed at the National High Magnetic Field Laboratory at the Florida State University (Figure 5.1).

Induction furnace heating relies on the production of eddy currents in a conducting work piece via inductive coupling between an oscillating magnetic field and conduction electrons. At the most basic level, this results in resistance heating that is accomplished by driving an electrical current in an electrically conducting solid. This is done by winding a coil (e.g., as a solenoid) around a work piece (idealized as a cylinder) and then driving an oscillating electrical current in the coil to produce an oscillating magnetic field that drives

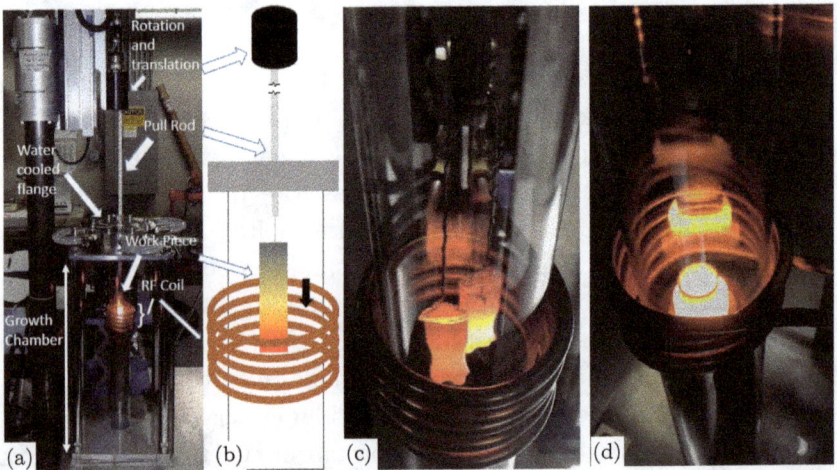

Figure 5.1. (a) Example of a custom-built crystal growth apparatus showing the inert gas growth chamber, RF-heating coil, tantalum ampoule (work piece), alumina pull rod, and linear translator and rotation motor. (b) Schematic of the crystal growth apparatus. (c) Photograph of the growth chamber with a sealed tantalum tube suspended from a braided tantalum wire inside of the RF coil. (d) Photograph of the growth chamber where an alumina crucible is placed inside of a graphite crucible. For (c) and (d), the applied RF field is coupled to the tantalum crucible and the graphite crucible, respectively.

electrons in the work piece. The resulting current distribution in the work piece is determined by the skin depth equation $I(r) = I_0 \exp(-r/d)$, where $I(r)$ is the current at a distance r from the surface, I_0 is the current at the surface, and d is the skin depth. A practical formula for skin depth in millimeters is given by the expression $d \propto (\rho/f\mu)^{1/2}$, where ρ is the electrical resistivity in Ωm, f is the frequency in Hertz and μ is the effective permeability in H/m of the material [1–3]. From this expression, it is seen that, for high conductivity metals (e.g., Ta or Nb) that are heated using an induction furnace operating at a typical frequency (hundreds of kHz), the penetration depth will be on the order of hundreds of microns. Since the active heating occurs in the skin layer, the bulk of the sample is heated through heat conduction to its core. This means that induction furnace heating occurs due to a non-trivial combination of eddy current heating, thermal heat conduction, and heat radiation to the environment. The result is that design of crystal growth equipment relies upon a large store of empirical information that has been gained from experience in the laboratory research setting and in industry. Several useful sources are given in [1–5] and many others can be found in the literature that give detailed technical discussions of core components such as coil design, dynamic frequency matching and optimization, etc. Here, we mainly describe practical considerations for crystal growth for research purposes that were gained during our crystal growth activities.

5.2. Apparatus Details

The power supply that was used for the experiments described below is an Ambrell 6 kW Easy Heat system. Its assembly and configuration is detailed in service notes that are associated with it. Systems with significantly higher power delivery are commercially available, and might be needed for other sample growth configurations. Besides Ambrell, there are several other companies that manufacture comparable power supplies including TRUMPF Huttinger, Pillar Induction, and MTI corporation. Besides the power supply, the most important consideration is sample environment, which must be tailored to each material and for the growth method that will be used. This involves design of the sample chamber and crucible selection.

In principle, the simplest possible configuration does not involve a sample chamber at all. Depending on the properties of a material, it may be possible to carry out a reaction in air or inside of a sealed quartz ampoule. One useful configuration is to wrap a standard alumina crucible and with a refractory metal susceptor (e.g., made from tantalum or niobium foil) and to seal them under vacuum inside of a quartz tube in order to protect the reactants from interacting with air at high temperatures. This simple arrangement is often adequate, but there are many operations that demand a more carefully designed environment. As shown in Figure 5.1, one approach is to make use of an enclosed tube where either a vacuum or an inert gas environment can be established. Here, the chambers are transparent quartz rods with diameters near 5.8 cm, around which the induction work coil is wound. The quartz tube is adapted to stainless steel flanges with rubber o-rings which feature several ports where either a vacuum pump is attached or custom gas is introduced to the chamber. For metallic melts, it is often sufficient to flow purified inert gas during a growth to protect the sample from oxidization. The chamber diameter is an important parameter to consider because it defines the smallest possible diameter of the work coil. Note that the energy density that can be delivered to a work piece depends on the fraction of the area inside the coil that is filled. In order to optimize the power transfer it is important to maximize this number, but this has to be balanced against practical considerations relating to how a work piece will be loaded into and occupy the space. Note that an alternative approach to establish an ultra-high vacuum environment is described in [4, 5].

The sample environment can then be customized for different growth processes. The simplest configuration is one where the work piece is placed directly in the center of the coil and remains static in that position. This is typically accomplished either by placing it on top of a pedestal or by suspending it from wire. In both configurations, it is required to consider the properties of the wire or pedestal at high temperatures, under rapid heating, and at large temperature gradients. For instance, it is tempting to consider alumina posts or crucibles as useful pedestal materials because they are common resources in most crystal growth laboratories. However, the strain that is induced under a large temperature gradient and at high temperature frequently causes this material to shatter. A much more robust material for such operations is boron nitride, which features good thermal shock resistance [1–3]. If the sample is suspended from a wire, it is simple to use a refractory metal (e.g., W, Mo, Ta, or Nb) which can be braided to create a durable fiber that will withstand being exposed to high temperatures for hours or even days, so long as it does not oxidize.

Beyond the static sample configuration, it may also be useful to either translate the sample or the hot zone: e.g., as is done for the Bridgman and molten metal flux (see below) techniques. As shown in Figure 5.1, this is accomplished (1) by attaching the sample to a pedestal or wire that is mounted to a linear translator or (2) by mounting the coil work head to a linear translator. It is also straightforward to heat a workpiece in a static configuration to establish a molten melt and then to insert a refractory metal pull-rod into it and subsequently pull a crystal from the melt using the Czochralski technique (see below).

As with other crystal growth methods, crucible selection is key. The first priority is to select a crucible material that does not react with the material that is being heated. In practice, this may require a process of "guess and check" that is guided by close inspection of binary phase diagrams. Possible crucible materials include W, Mo, Nb, Ta, Pt, graphite, alumina, boron nitride, although there are many others (Table 5.1 and [1–3]). Metallic crucibles are often convenient because the RF-field directly heats them by coupling directly to their conduction electrons. Furthermore, for malleable metals (often Nb, Ta, Pt) it is straightforward to produce sealed tubes using commercially available tube stock. As shown in Figure 5.2(d), one approach is to weld the ends closed using an electrical arc. This approach is particularly useful when a

Table 5.1.   Summary of some crucible materials and their highest useful temperatures ([1–3]).

Material	Maximum temperature (°C)	Melting temperature (°C)
Tungsten	3000	3422
Tantalum	2700	3017
Graphite	2600	3600
Aluminum oxide	1900	2072
Boron nitride	1700	2973
Platinum	1600	1768
Nickel	1300	1455
Silica	1250	1710
Pyrex	500	820

Note that these values are below the material's melting temperature, and are meant to provide a rough guide to their range of usefulness.

reaction contains elements that have high vapor pressures that can be contained inside of a sealed tube. While insulating crucibles such as those made from aluminum oxide or boron nitride do not couple to the RF field, they can be heated by wrapping them in a refractory metal susceptor or a graphite sheath (Figures 5.2(a) and 5.2(c)). For some operations, it is useful to build a reusable water cooled metal hearth. Finally, under an optimized configuration, it is even possible to magnetically levitate the sample during a growth. This is desirable to avoid introduction of impurities from the crucible to the sample or when the sample itself is aggressively reactive with all possible crucibles (e.g., as for $YbNi_4P_2$ [6]).

Lastly, it is important to consider the coil dimensions and its position with respect to the work piece. Many modern power supplies are equipped with a mobile work-head, where the coil itself is attached. Custom coils can be wound using standard soft copper alloy tubing, which allows for flexibility in designing their diameter and length. Many other approaches to coil production are also used. There are two general considerations in designing a coil: (1) a smaller diameter results in a larger energy density (i.e., higher temperatures are accessible) and (2) as length increases the temperature gradient that will be produced along a work piece will be more gradual. Both of these factors contribute to the determination of where to

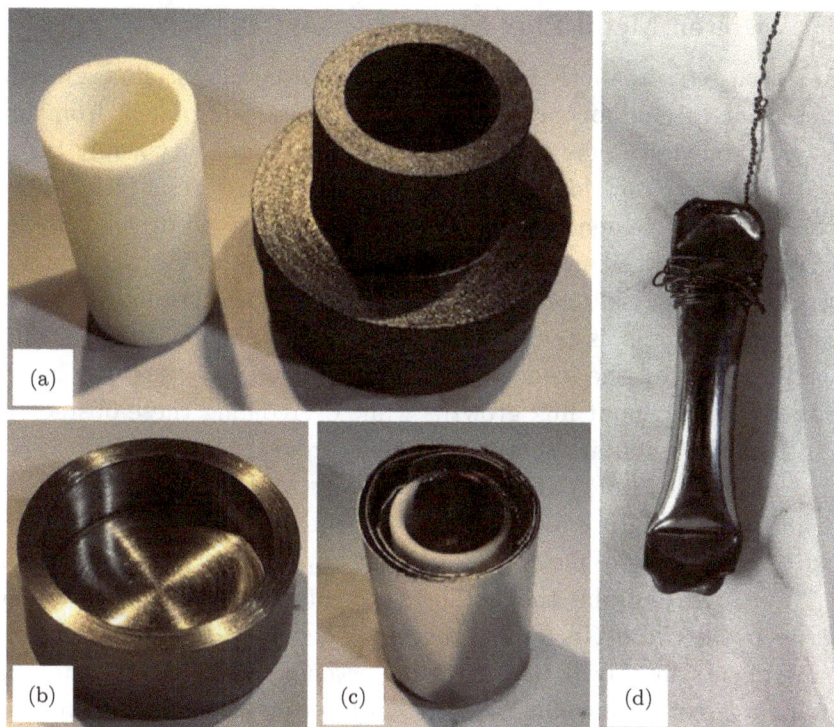

Figure 5.2. Photographs of typical crucible materials and susceptor sheaths that readily couple to an applied RF field. (a) A custom graphite crucible that has been machined so that its inner diameter matches the outer diameter of the non-conducting 2mL alumina crucible. (b) Custom fabricated tungsten crucible with inner diameter 0.16 cm, outer diameter 0.21 cm, and height 0.31 cm. (c) 2 mL alumina crucible that is wrapped in tantalum foil, which acts as a susceptor that couples to an applied RF field. (d) Tantalum tube that has been sealed using an arc furnace with reactants inside of it.

place the coil with respect to the sample chamber. For instance, as shown in Figure 5.1, the coil diameter may be limited by the quartz tube diameter of a sample chamber. On the other hand, if a higher energy density is required, the coil may be placed inside of the tube and directly around the sample. This has the advantage of allowing access to higher temperatures but in this configuration the coil will be exposed to whatever material vaporizes during the reaction.

5.3. Molten Metal Flux Method

As discussed in another chapter, the molten metal flux growth method is useful for producing single crystal specimens from an excess (non-stoichiometric) melt [8–11]. This process typically takes place inside of a closed furnace that is heated resistively. This approach has tremendous utility, but also has some drawbacks. In particular, growths are time consuming (usually lasting weeks), the stirring of the melt occurs mainly due to slow thermally driven convection, and it is not easy to observe the reaction space during the growth. In the context of induction furnace heating, it is possible to carry out flux growths while overcoming these obstacles. Particularly useful are that the RF field provides active stirring and that optical access (1) allows for knowledge about ampoule integrity or breakdown during a growth and (2) may even enable additional active monitoring of melts: e.g., using X-ray and scattering methods.

Here we describe the details for synthesis of UPt_2Si_2, which forms in the $CaBe_2Ge_2$-type structure. The apparatus used for these experiments in shown in Figure 5.1. Elemental U, Pt, Si, and In (purities greater than 99.95%) were cut into small pieces, combined in the molar ratio 1:2:2:40, and sealed inside a tantalum crucible under argon gas using a standard arc-furnace. This is similar to what was reported previously for molten metal flux growth of URu_2Si_2 [12], although here the molar percentage of indium was increased. The tantalum ampoule was formed from commercially available tubing with an outer diameter of 1.01 cm, an inner diameter of 0.76 cm, and was cut to a length of 5 cm. Once sealed, the ampoule was vertically suspended inside the growth chamber from a braided tantalum wire that was attached to an alumina pull rod. A growth environment of purified Ar gas was maintained near one atmosphere during the growth. Heating was accomplished using a two-turn coil that was wrapped around the quartz tube and was controlled by a 6 kW power supply. During the growth, the temperature T_M was measured using a pyrometer, where the output reading is roughly 10% less than the actual sample temperature, due to loss of light intensity as it travels through the quartz sample chamber. The sample temperature was rapidly raised to $T_M = 1235°C$ (sample temperature $T_S \approx 1350°C$)

and allowed to dwell here for four hours. The sample was then cooled to $T_M = 990°C$ over 21 hours, after which it was fast cooled to $T_M = 730°C$ over one hour and then rapidly quenched to room temperature by turning off the power supply. Finally, the ampoules were opened and the indium was removed by etching the resulting ingot in hydrochloric acid.

Pictures of UPt$_2$Si$_2$ crystals, transmission electron microscopy scans, and representative magnetic susceptibility χ vs. temperature and electrical resistivity ρ vs. temperature curves are shown in Figure 5.3. These measurements show that these specimens crystallize in the expected crystalline structure and exhibit behavior that

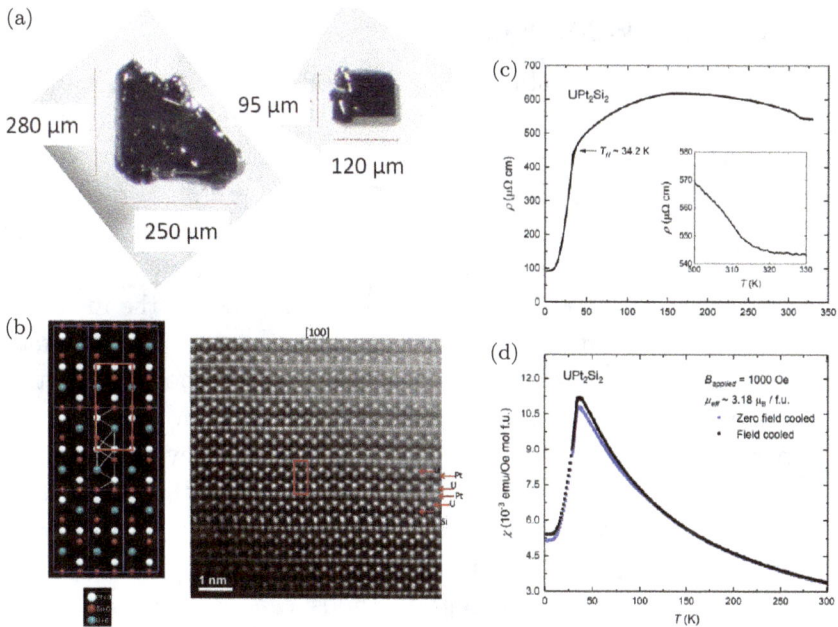

Figure 5.3.   Summary of results for UPt$_2$Si$_2$ single crystals that were produced using the molten metal flux growth technique described in the main text, where RF heating was used. (a) Representative single crystals of UPt$_2$Si$_2$ that were produced using this technique. (b) Schematic representation of the crystal structure and scanning tunneling electron microscopy results showing that this structure is present in these crystals. (c) Electrical resistivity ρ vs. temperature T for a typical specimen. (d) Magnetic susceptibility $\chi(T)$ for an aligned mosaic of typical specimens. Results shown in (c) and (d) are consistent with earlier results for this compound [13].

is comparable to what has been seen previously for single crystals produced using the Czochralski technique [13]. It is noteworthy that although this process takes significantly less time than a typical flux growth in a resistive furnace, the resulting crystals have high quality and similar size to what we are otherwise able to produce. We speculate that this is due to the active stirring that is provided by the RF field, which may decrease the growth time. Thus, this method has the potential to significantly accelerate experiments that require production of large numbers of batches: e.g., chemical substitution series or exploration of the chemical phase space to uncover new materials.

5.4. Czochralski Method

The Czochralski or crystal pulling method is a common means to produce large single crystal specimens [1–4], and is most useful for materials that form directly upon solidification from a melt (congruently melting). In general terms, the approach is to prepare a reservoir of liquid material that is maintained at temperatures just above the melting point for a given mixture of elements. A seed crystal (or other pull rod material) is then dipped into the melt and gradually withdrawn to produce the ingot. In order to encourage mixing, the pull rod is often rotated. Especially important for this method is that the pull and rotation rates be smooth, that the melt temperature be well controlled, that the melt not evaporate, that the crucible be inert to the melt, and that the growth environment (vacuum or gas) not contaminate the melt. In practice, a successful growth is likely to be a compromise between optimization of all of these factors. Many different methods can be used to heat a melt including RF heating, resistive heating, lasers, focused light, electron beams, and arcs. The most common approaches are RF and arc heating, and here we focus on the former. The advantages of RF heating over arc melting include that the temperature can be gradually increased, allowing for a precise determination of the melting temperature and that the heat is uniformly distributed and that certain configurations (e.g. Hukin-type crucible) allow levitation of the melt.

As for all melt growths, the crucible choice is of high importance. Several useful crucibles are listed in Table 5.1 along with their melting temperatures. Most of these materials are useful because they have high melting temperatures, are generically inert, can be obtained in a high purity form, have high strength and are durable under temperature gradients, and have low porosity. Ideally, a useful crucible will also be easily cleaned (e.g., using acid) and is machinable. Additionally, it is always necessary to consider whether they will be dissolved by a specific melt: some insight is gained by consulting binary phase diagrams, but in practice each growth experiment may present specific challenges that were not expected. While some of these materials are metallic, and thus couple directly to the RF field, others are non-conducting. As described above, it is straightforward to wrap such crucibles in a susceptor material such as Ta or Nb foil or a graphite tube.

A second consideration when using RF heating for Czochralski growth is to ensure that the pull rod does not heat to the temperature of the melt or dissolve into the melt. The most straightforward way to do this is to build the rod from a high strength non-conducting refractory material such as an alumina rod. As shown in Figure 5.4, an alumina rod is easily attached to a rotation/translation stage and introduced into a growth chamber through a sliding seal. The alumina may be sharpened and inserted directly into the melt, or a stinger tip (e.g., made from braided Ta wire) may be attached to the alumina. More sophisticated designs involve a seed crystal clamp that is affixed to the end of the pull rod.

Here we describe the details for synthesis of $A\mathrm{Pd}_3$ (A = Sn and Pb), which were obtained through the Czochralski process (Figure 5.4). For both growths, the melts were prepared by combining the pure elements in a molar ratio of A:Pd = 1:3. For both of these compounds, inspection of the binary phase diagrams suggested that tungsten crucibles would be robust against chemical reaction with the melt and that the desired phase forms directly from the melt as a line compound. As shown in Figure 5.4, cylindrical W crucibles with dimensions inner diameter 0.16 cm, outer diameter 0.21 cm, and height 0.31 cm were filled with total charge masses of 20 g and 25 g for SnPd_3 and PbPd_3, respectively. The crucibles were place on top

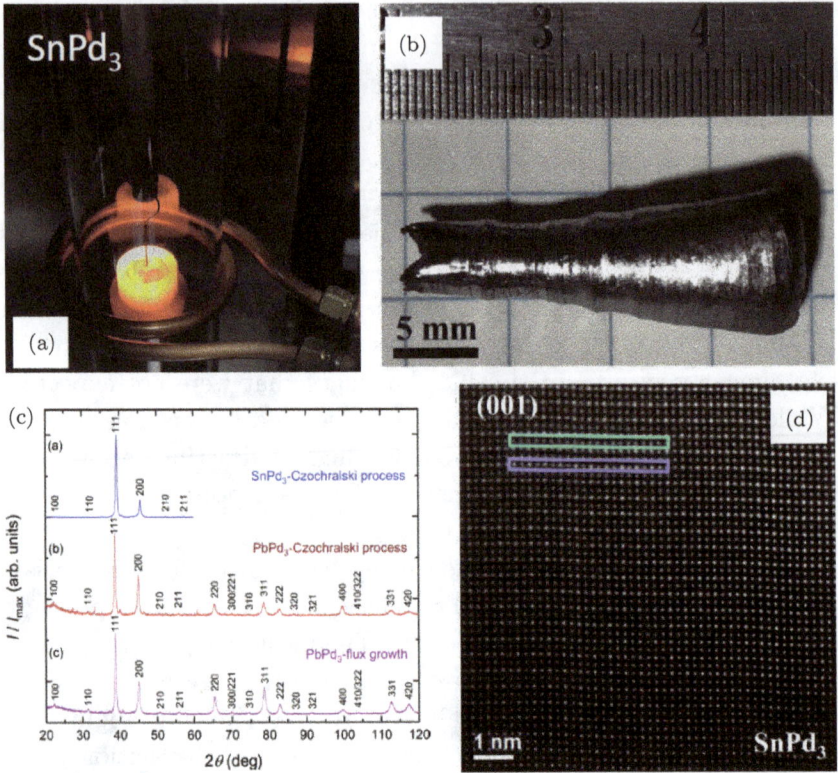

Figure 5.4.   (a) Photo of crystal growth apparatus where it is in use pulling a single crystal of SnPd$_3$ using the Czochralski method. Here, the alumina pull rod is tipped using a tantalum wire and the unmelted pre-reacted polycrystalline SnPd$_3$ is seen as well-defined boules in the heated tungsten crucible. (b) Single crystal specimen that was pulled from the melt. (c) Powder X-ray diffraction measurements were collected for SnPd$_3$ and PbPd$_3$ crystals that were produced using this method. Also shown for comparison is a powder X-ray diffraction scan for PbPd$_3$ single crystals that were produced using the molten metal flux growth technique [14]. (d) HR-TEM image of SnPd$_3$, indicating the system is well ordered.

of boron nitride pillars during the growth, where the weight of the crucible + sample was sufficient to prevent movement during pulling. Upon heating, it was observed that the crucible was first heated through coupling to the rf-field, after which heat was transferred to the A:Pd mixture, whose temperature lagged behind that of the crucible throughout the entire growth. The measured temperatures

T_M where a melt was established were 1200°C and 1350°C for PbPd$_3$ and SnPd$_3$, respectively. It is important to reiterate that T_M was measured using a pyrometer, where the output reading is roughly 10% less than the actual sample temperature, due to loss of light intensity as it travels through the quartz sample chamber. Once a liquid melt was established, the tip of the pull rod was inserted into the melt to establish a seed crystal. The rod was then extracted from the melt with a pulling at a speed of 5 mm/h. As is often the case during growths of this sort, it was necessary to dynamically adjust the power in order to maintain the melt and control the crystal diameter as the sample is extracted from the melt. The growth was carried out over a 10 hour period to produce specimens with lengths near 2.5 cm and diameter near 1 cm.

Pictures of the APd$_3$ crystals and transmission electron microscopy scans are shown in Figure 5.4. Following the growth, the crystals were sealed under vacuum in quartz tubes and annealed at 800°C and 950°C for PbPd$_3$ and SnPd$_3$, respectively, for 30 days to release any strain formed during the Czochralski process. A comparison was also made to specimens grown using the molten metal flux growth technique, which revealed little difference. Measurements performed on these crystals revealed quantum oscillations, showing that high quality specimens are readily produced using this method. These results and other electronic properties measurements were reported in [14].

5.5. Conclusions

It seems appropriate to conclude this chapter by propagating forward a quote from W. G. Pfann from 1967 that was made specifically regarding zone melting, but has broader relevance for contemporary crystal growth experimental efforts:

> *I regard the conception and development of zone melting as an exciting scientific advance. And I cannot help being saddened by to hear it occasionally referred to as simply a technical innovation that was mysteriously evoked by the need for transistor grade germanium and silicon. I regard zone melting as elegant both in its simplicity*

and its surprising complexity. I also regard it to this day as a wonderful adventure, filled with surprise and joy.

Roughly 50 years later, we find experimental research into crystal growth methods in much the same position. This is despite the regularity with which discoveries of new materials have spurred progress in physics, chemistry, and materials science. Therefore, it is hoped that the information presented here not only provides practical guidance into the creation and use of crystal growth equipment based on the RF heating technique, but also encourages research in this stimulating field that lies at the intersection between basic and applied science.

Acknowledgments

This work was performed at the National High Magnetic Field Laboratory (NHMFL), which is supported by National Science Foundation Cooperative Agreement No. DMR-1157490 and DMR-1644779, and the State of Florida. Research addressing actinide materials was supported by the Center for Actinide Science and Technology, an Energy Frontier Research Center funded by the U.S. Department of Energy (DOE), Office of Science, Basic Energy Sciences (BES), under Award Number DE-SC0016568. Synthesis studies addressing UPt_2Si_2 were performed by Greta Chappell, Yan Xin, and R. E. Baumbach. Chappell was supported as a graduate student by the NHMFL. Synthesis studies addressing $(Sn,Pb)Pd_3$ were performed by Kaya Wei and R. E. Baumbach. Kaya Wei was supported by the NHMFL through the Jack E. Crow Postdoctoral Fellowship. Finally, R. E. Baumbach extends his thanks to the Editors of this book, for their tremendous patience while this chapter was being written.

Bibliography

[1] J. S. Shah, in B. R. Pamplin (ed.), *Crystal Growth* (1975), Pergamon Press Ltd., pp. 104–156.
[2] J. S. Shah, in B. R. Pamplin (ed.), *Crystal Growth* (1975), Pergamon Press Ltd., pp. 326–396.

[3] R. C. J. Draper, in B. R. Pamplin (ed.), *Crystal Growth* (1975), Pergamon Press Ltd., pp. 497–520.

[4] A. Bauer, G. Benka, A. Regnat, C. Franz, and C. Pfleiderer, *Rev. Sci. Instrum.*, **87**, 113902 (2016).

[5] A. Bauer, A. Neubauer, W. Munzer, A. Regnat, G. Benka, M. Meven, B. Pedersen, and C. Pfleiderer, *Rev. Sci. Instrum.*, **87**, 063909 (2016).

[6] K. Kliemt, and C. Krellner, *J. Crystal Growth*, **449**, 129 (2016).

[7] T. Onimaru, Y. F. Inoue, A. Ishida, K. Umeo, Y. Oohara, T. J. Sato, D. T. Adroja, and T. Takabatake, *J. Phys. Conden. Matter*, **31**, 125603 (2019).

[8] Z. Fisk, and J. P. Remeika, *Growth of Single Crystals from Molten Metal Fluxes, Handbook on the Physics and Chemistry of Rare Earths, Vol. 12*, K. A. Gschneidner Jr and L. Eyring (eds.) (Elsevier Science, 1989).

[9] P. C. Canfield, and Z. Fisk, *Philos. Mag. B*, **65**, 1117 (1991).

[10] M. G. Kanatzidis, R. Pottgen, and W. Jeitschko, *Angew. Chem.*, **44**, 6996 (2005).

[11] W. A. Phelan, M. C. Menard, M. J. Kangas, G. T. McCandless, B. L. Drake, and J. Y. Chan, *Chem. Mater.*, **24**, 409 (2012).

[12] R. E. Baumbach, Z. Fisk, F. Ronning, R. Movshovich, J. D. Thompson, and E. D. Bauer, *Philos. Mag.*, **94**, 3663 (2014).

[13] M. Bleckmann, A. Otop, S. Sullow, R. Feyerherm, J. Klenke, A. Loose, R. W. A. Hendrikx, J. A. Mydosh, and H. Amitsuka, *J. Magn. Magn. Mater.*, **322**, 2447 (2010).

[14] K. Wei, K.-W. Chen, J. N. Neu, Y. Lai, G. L. Chappell, G. S. Nolas, D. E. Graf, Y. Xin, L. Balicas, R. E. Baumbach, and T. Siegrist, *Phys. Rev. Mater.*, **3**, 041201 (2019).

Chapter 6

Hydrothermal Synthesis and Crystal Growth

Brandon Wilfong[*,‡], Xiuquan Zhou[†,§], and Efrain E. Rodriguez[*,‖]

*Department of Chemistry and Biochemistry,
University of Maryland, College Park, MD 20742, USA
†Materials Science Division, Argonne National Laboratory,
Lemont, IL 60439, USA
‡bwilfong@umd.edu
§xzhou1@terpmail.umd.edu
‖efrain@umd.edu

Everything is made of something. If you don't have the material technology, you lose the device, the system, the application.

— Robert A. Laudise, AT&T Bell Laboratories

6.1. The Hydrothermal Generation of Inorganic Solids

In the ocean's ridges, openings near volcanic activity heat seawater above the boiling point of pure water and even up to 400°C [1]. These hot vents generate mineral ores typically insoluble in water, from iron sulfides to manganese oxides. Known as hydrothermal vents, these natural chemical reactors eject dissolved inorganic species into the frigid ocean waters, depositing in the process the minerals key to aquatic ecosystems [2]. In the laboratory, we can reproduce similar conditions found in nature to prepare especially difficult inorganic solids. Therefore, it is no coincidence that the hydrothermal method was first developed in the geological and geochemical sciences [3]. The technique has matured enough that it is now the purview of chemists, physicists, engineers and materials scientists involved in crystal growth.

In this chapter, we briefly introduce the reader to the principles of hydrothermal synthesis, crystal growth and processing. We end the chapter with specific examples in the area of quantum materials. While the technique remains underdeveloped for the preparation of high-quality quantum materials, the hydrothermal technique could gain in prominence in this field for the following reasons: (1) it remains a relatively low-energy technique compared to energy intensive methods such as image furnace growth; (2) it can be relatively cost effective since it only requires a high-pressure vessel and an oven; (3) the slow growth of high-quality crystals from liquid solutions remains an attractive possibility since higher temperature methods such as molten fluxes may impart more defects, inclusions, and impurities in the final product; and (4) some metastable phases may only be prepared at relatively low temperatures ($<300°C$), which preclude conventional solid-state syntheses that require high temperatures for diffusion and homogeneous mixing. All of these advantages make hydrothermal suitable not only for industrial applications but also for academic research dedicated to the synthesis of new materials [4].

The hydrothermal method will obviously not work for every inorganic system, especially if the final product is water sensitive. However, workers in this field have managed to grow a wide variety of solids. Simple and complex metal oxides [5, 6] are sulfides, selenides, halides, phosphates, carbonates, molybdates, and silicates [3]. Instead of synthesizing compounds of the elements, one can hydrothermally isolate native metals such as gold and aluminum by changing the reaction conditions. Figure 6.1 shows the diversity of structure types and compounds that have been successfully prepared through the hydrothermal and related solvothermal methods.

Not all the materials are purely inorganic or dense bulk crystals. For example, microporous aluminosilicates known as zelolite can be easily prepared in this manner as can metal oxides with larger pores known as mesoporous materials [7, 8]. The hydrothermal method and the closely related solvothermal method have been reliable routes for the crystal growth of metal-organic frameworks (MOFs) [9],

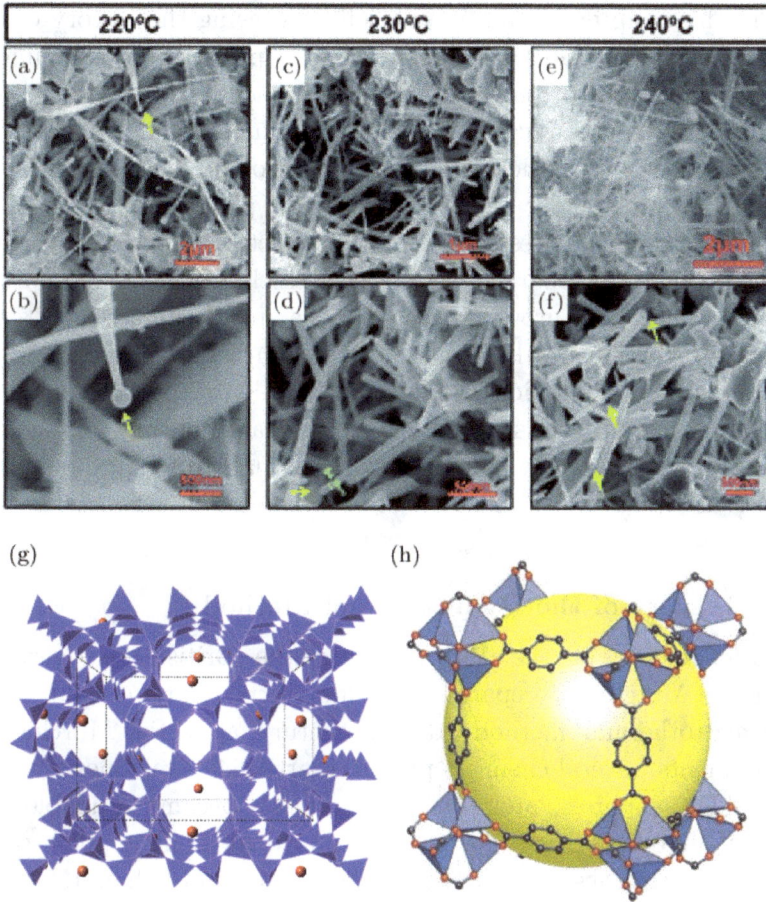

Figure 6.1.   Examples of materials prepared by the hydrothermal/solvothermal treatment. (a–f) Microscopy images of nanowires of SmB_6, a topological insulator, grown the hydrothermal method at various temperatures [11]. (g) The microporous zeolite ZSM-5 and (h) the metal-organic framework MOF-5 [12].

which are comprised of inorganic units (e.g., metal oxide polyhedra) connected by organic linkers such as carboxylates and imidazolates. Finally, one can grow nanomaterials with prescribed shapes such as wires or platelets, or nanocrystals with only particular crystal facets exposed [10]. The case is therefore easily made that hydrothermal will remain an important tool in the repertoire of quantum materials syntheses.

In this chapter, we start by briefly reviewing the history of the hydrothermal method followed by a description of water's unique properties as a solvent both under ambient and hydrothermal conditions. We then cover the basic thermodynamic and kinetic concepts useful for understanding crystal growth; in this section, we introduce to the reader the different types of phase diagrams used by the community. We present some of the practical aspects concerning the hydrothermal apparatus such as the vessels and liners appropriate for different crystal growths. Finally, we end this chapter with some recent applications in the preparation of iron-based superconductors and quantum spin liquids. While the field of hydrothermal chemistry and materials synthesis is well established and mature, we suggest some future directions for the hydrothermal growth of quantum materials.

6.2. History of the Hydrothermal Method

In the middle of the 19th century, the British geologist Sir Roderich Murchison prepared the first artificial quartz crystals using the hydrothermal method [13]. A century after Sir Murchison's experiments, several chemists produced artificial gemstones. Rubies, emeralds, sapphires, and over 130 minerals became available due to the discovery that water heated above its boiling point could dissolve gemstones or their precursors. After more than 150 years of development, the hydrothermal technique remains relevant for the dissolution, synthesis, processing, and single crystal growth of inorganic solids. Where it might have taken Earth thousands of years to produce a rare gemstone, the laboratory apparatus may take as long as a few days to hydrothermally produce a crystal of high quality.

The hydrothermal method was first focused on silicate materials and by the middle of the 20th century became key for the synthesis of aluminosilicates and related microporous zeolites [3]. First, hydrothermal synthesis was nearly synonymous with the preparation of silicates, and most reactions took place in sealed glass tubing. The products were usually fine powders, and since powder diffraction techniques were not yet developed, the hydrothermal method remained largely a niche tool with limited utility. The advent of better hydrothermal apparatus was key towards its development as a crystal growth and materials synthesis technique. For this purpose, workers mostly in France, Germany, and other parts of Europe developed the Morey autoclave and test-tube autoclaves, which allowed them to achieve sufficiently high pressures and temperatures [3]. They extended the synthesis of pure silicas to ternary silicates including clays, zeolites, and other chemically related systems. Finally, the advent of the cold-seal autoclave such as the Tuttle-cone seal apparatus led to a popularization of hydrothermal research (Figure 6.2(a)). Not only were silica-based crystals reliably grown

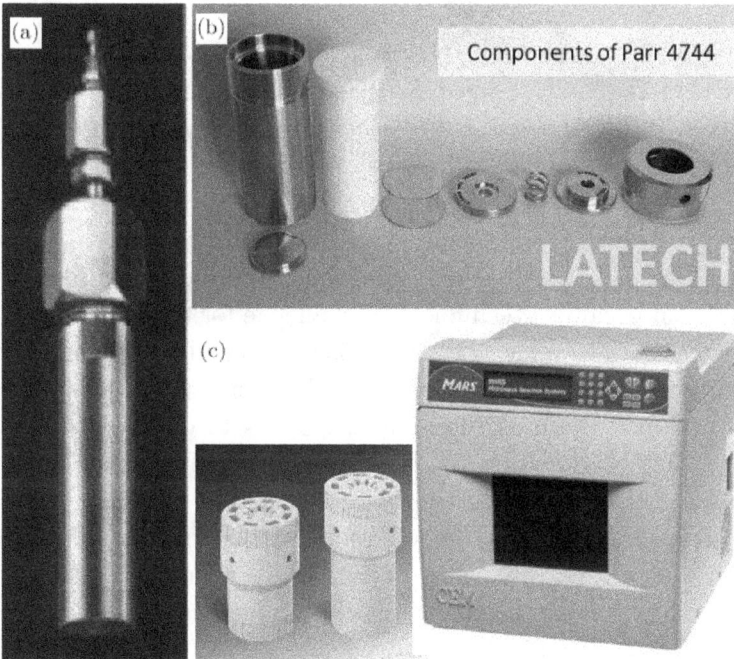

Figure 6.2. Examples of modern autoclaves used for hydrothermal synthesis. (a) The Tuttle cone sealed stainless steel autoclave that became available in the middle of the 20th century [3]. (b) The typical Parr autoclaves used with components shown such as Teflon liner cup, rupture disks, and spring [9]. (c) All Teflon autoclaves utilized in microwave heating for fast, high-throughput hydrothermal synthesis [10].

now but a host of natural minerals including corundum and emeralds. Later, the Parr autoclaves (also known as the acid digestion bombs) became popular for the synthesis of zeolites, MOFs, nanocrystalline materials, etc. (Figure 6.2(b)). One of the newest innovations has been the all-Teflon autoclave that can safely go into microwave reactors for fast hydrothermal synthesis (Figure 6.2(c)).

For the decades spanning 1950–1970, the hydrothermal technique became a maturing science as researchers investigated the thermodynamic and kinetic conditions under which materials formed. The first relevant $P - T$ and $P - V - T$ phase diagrams were formulated, and the method's progress became of international interest with major participation by the USSR, the US, and Japan in addition to Europe [3]. Not only were natural minerals being synthesized but now a host of materials with no natural counterparts as well. This time was especially important for the synthesis of zeolites [7, 8]; these materials remain indispensable for the chemical industry as cracking catalysts, molecular sieves, and other applications arising from their finely tuned pore architectures.

An important historical lesson of how hydrothermal crystal growth advanced technologies can be found in the early 1960s. After WWII, a shortage of high-quality quartz crystals imported from Brazil became a technological challenge for telecommunications companies such as AT&T. Used as oscillators and filters, high quality quartz became important for materials scientists to grow. Nacken first developed the hydrothermal techniques to grow crystals larger than a few centimeters [3]. Laudise at AT&T Bell Laboratories and others then produced even larger quartz crystals on an industrial scale. Figure 6.3(a) shows racks of quartz grown from seed crystals suspended into large hydrothermal furnaces buried in the ground. The largest producer became Japan using this technique (Figure 6.3(b)).

A lull and decline in the science of hydrothermal chemistry occurred during the 1980s. While there are likely a combination of geopolitical reasons for this [3], important technological ones include the development of methods such as molecular chemical vapor deposition, chemical vapor transport, among others used in the

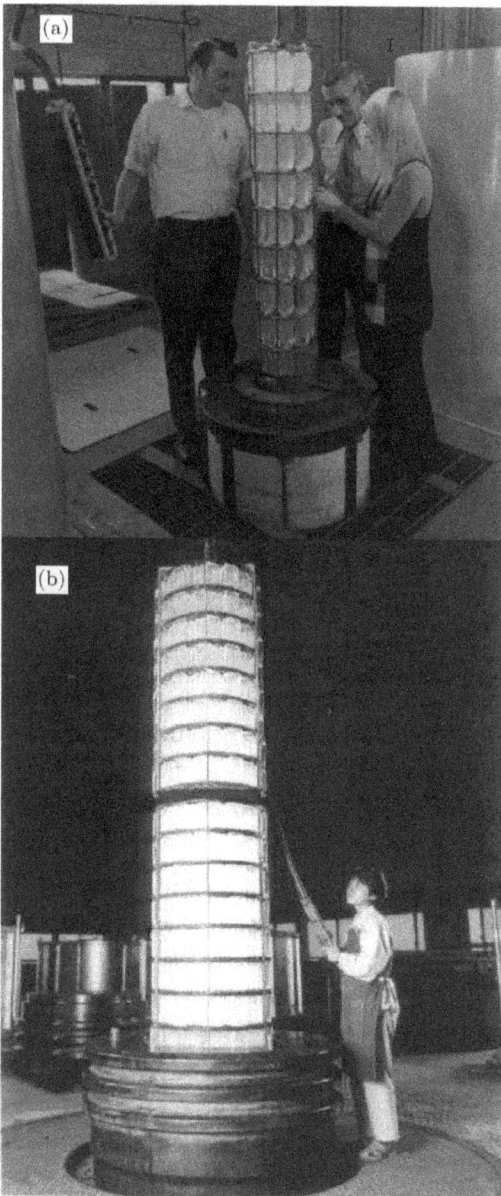

Figure 6.3.   Industrial scale hydrothermal growth of quartz. (a) Large underground autoclaves to grow racks of high-quality quartz crystals using a seeded growth technique. Picture from AT&T Bell Laboratories. (b) The largest quartz crystals for industrial use were produced in Japan by the Toyo Telecommunications Co. Photos from [3].

semiconductor industry. The technique made a comeback, however, once it was found that it was highly useful for the synthesis of MOFs (or coordination polymers), inorganic–organic hybrid materials, and nanomaterials. The number of papers with the word "hydrothermal" included in the title has grown since the beginning of the 1990s, and most of these papers are no longer in the geosciences but rather in the categories of materials science, physical chemistry, multidisciplinary chemical sciences, and applied physics.

6.3. Water: The Universal Solvent

What special conditions does the hydrothermal method entail? If we return to the image of hydrothermal vents on the ocean floor, we can find what makes this technique unique from other crystal growth methods. First, water must be heated above its boiling point, so that the vapor pressure above the liquid line must be greater than 1 atm. These high-pressure conditions can be met at the bottom of the ocean floors. The original definition of hydrothermal reaction included raising water above its critical point, which is approximately 374°C for pure water. However, the definition has been extended to include much milder conditions. *We define hydrothermal conditions as any heterogeneous reaction in water above 100° C and at pressures above 1 atm.* Since the critical point of water increases for number of solute species dissolved and for higher pressures, it is difficult to even know what the critical point of an aqueous solution under hydrothermal conditions. Even at 800°C, under sufficient pressure the critical point may not be surpassed.

Another typical aspect of hydrothermal synthesis is that the aqueous solution includes an additive called a mineralizer. In the case of the natural hydrothermal plumes, the seawater contains plenty of dissolved salts that act as the mineralizers. In the laboratory, the mineralizer can consist of a simple salt such as NaCl, a base such as a metal hydroxide, or an acid such as HCl. As shown by Raoult's and Henry's Laws, the colligative properties of solutions can be drastically altered by the amount of solute (and therefore mineralizer) present. In these syntheses, the major parameters to optimize for product selectivity and crystal growth will include concentration of the reagents (organic or inorganic), pH of the solution, reaction time, soak temperature, and temperature profile across the apparatus.

Figure 6.4 gathers three phase diagrams that explain how the thermodynamic properties of water change during the hydrothermal process. The first diagram (Figure 6.4(a)) shows how the density of water changes as a function of temperature and pressure [15]. At pressures below 1 kbar, the density drops sharply at elevated temperatures. In fact, for 100 bars and 500°C, the density of water will diminish to 10% of its value under ambient conditions (approximately 1 g· cc^{-1}). The importance of this change in density is that the viscosity of water decreases under typical hydrothermal conditions. This change in density aids crystal growth since mass transport of ions and other solute species increases upon decreasing the viscosity of the aqueous solution.

Figure 6.4(b) also presents how the dielectric constant of pure water changes as a function of temperature and pressure [15]. The most important feature of this diagram is that temperature is a higher consideration for controlling the dielectric constant than pressure. The third thermodynamic diagram is one of the most important for anyone running a hydrothermal reaction. The diagram in Figure 6.4(c) shows the pressures that one will reach at a given temperature based on the percent filling of the autoclave's volume with pure water [16]. Interestingly, one sees that as long as the percent filling remains below 80%, then the risk of dangerously over-pressurizing the autoclave is minimized under the typical "soft" hydrothermal conditions where the temperature remains below 200°C.

While water is known as the universal solvent, there are certainly many cases where one would like to carry out hydrothermal reactions without water. In fact, this technique is known as solvothermal synthesis since it involves any other solvent besides water [16–18]. Briefly, one would like to use other polar solvents then typically one has a choice of inorganic and organic solvents. For example, to prepare nitrides, amides, and imides, ammonia is a good choice for the solvent. The literature sometimes refers to these NH_3 reactions

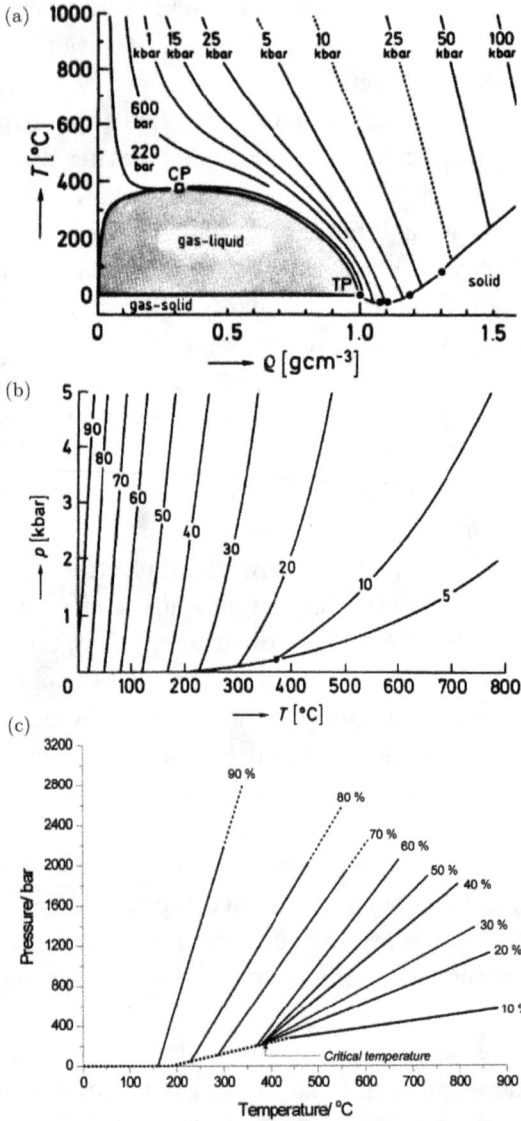

Figure 6.4. Thermodynamic diagrams relating properties such as pressure, temperature, volume, density, and dielectric constant for water. (a) Isobars show the relationship between density r and temperature. The triple point (TP) and critical point (CP) for pure water are also labeled. (b) Curves representing the dielectric constant at set pressures and temperatures. Figures from [15]. (c) Isochores for pressures in an hydrothermal autoclave with temperature. The percentage indicates percent filling of the reaction vessel with pure water. Figure 6.4(c) from [16].

Table 6.1. Relevant properties of solvents commonly used for solvothermal synthesis.

Solvent	Dielectric constant	Boiling point (°C)	Density $(g \cdot mL^{-1})$
Water	78.4	100.00	1.000
Ammonia (liquid)	17	−33.34	0.769
Ethanol	24.3	78.24	0.7893
Methanol	32.7	64.7	0.792
Dimethyl-sulfoxide (DMSO)	46.7	189	1.1004
Dimethyl-formamide (DMF)	36.7	152	0.948
Ethylene glycol	37.0	197.3	1.1132
Tetrahydro-furan (THF)	7.6	66	0.8892

Those of water included for comparison. These properties [19] are for pressures near 1 bar, and the dielectric constant for near room temperature or slightly below.

as ammonothermal synthesis. Alcohols such as methanol and ethanol are also popularly used in solvothermal synthesis. More recently, the preparation of MOFs and other porous materials has taken place in polar aprotic solvents. These include dimethyl sulfoxide (DMSO) and dimethylformamide (DMF). Finally, if one wants to use a solvent with a large boiling point, then ethylene glycol is a good choice. Table 6.1 gathers some important properties of water and the nonaqueous solvents for synthesis.

6.4. Thermodynamics of Aqueous Solutions Under Hydrothermal Conditions

The governing thermodynamic principles applying to hydrothermal growths are well-studied in their general case [20–24]. However, systematic study of the thermodynamics of complex mixtures at various concentrations, temperatures, pressures, and pH are often lacking in the hydrothermal literature. This is often why hydrothermal synthesis is considered a "black box" reaction scheme. However, in recent years several groups have worked to remedy this by using extensive thermodynamic data to model and intelligently design hydrothermal reactions to target desired phases [25–31]. It is still informative to work through a basic hydrothermal reaction from theoretical principles to illustrate how hydrothermal synthesis works on a fundamental level.

The theory of thermodynamics of mixtures, aqueous or otherwise, is very well detailed in elementary texts in thermodynamics [32–36]; however, the application to simple hydrothermal reaction schemes is enlightening. In general, it is desirable to approach the formalism through analysis of the use of the Gibb's free energy for the reaction. This is 2-fold: (1) chemists think of reaction in terms of Gibb's free energy, and (2) Gibb's free energy is a function of temperature, pressure and moles of compound (Eqs. (6.1) and (6.2)) which are typically the easiest variables to control in experimental setup:

$$G = f(T, P, n_i, n_j, \ldots, n_c), \tag{6.1}$$

$$dG = \left(\frac{\partial G}{\partial T}\right)_{P,n} dT + \left(\frac{\partial G}{\partial P}\right)_{T,n} dP + \sum_{i=1}^{C} \left(\frac{\partial G}{\partial n_i}\right)_{T,P} dn_i, \tag{6.2}$$

where n_i is the number of moles of the ith component in a C-component mixture including the solvent used in the system. An extreme simplification comes from the design of hydrothermal experiments. Typically, these experiments are done at constant temperature and pressure, at least in particular region of the growth apparatus. This simplifies the Gibb's free energy of the system to:

$$dG = \sum_{i=1}^{C} \left(\frac{\partial G}{\partial n_i}\right)_{T,P} dn_i, \tag{6.3}$$

$$dG = \sum_{i=1}^{C} \mu_i dn_i, \tag{6.4}$$

$$\mu_i \equiv \left(\frac{\partial G}{\partial n_i}\right)_{T,P}, \tag{6.5}$$

$$G = \sum_{i=1}^{C} \mu_i n_i. \tag{6.6}$$

Such that the Gibb's free energy of the reaction, or the driving force for any chemical change is just a function of the chemical potential, μ, of each component of the system. Now, consider the simplest hydrothermal setup with three components: n_s number of moles of solid, n_d number of moles of dissolved solid, and n_{solv} number of moles of solvent. This allows the description of the free energy of the system to be:

$$G = \mu_s n_s + \mu_d n_d + \mu_{\text{solv}} n_{\text{solv}}. \tag{6.7}$$

If the dissolution proceeds at a fixed temperature and pressure in the reaction vessel, and the moles of solvent does not change, incremental change in the free energy of the system can be described as

$$G + dG = \mu_s(n_s - dN) + \mu_d(n_d + dN) + \mu_{\text{solv}} n_{\text{solv}}, \tag{6.8}$$

$$dG = (\mu_d - \mu_s)dN. \tag{6.9}$$

This equation yields three cases that are essential for crystal growth in the hydrothermal method. From a chemical reaction point of view, if dG is negative, a spontaneous chemical reaction will take place:

(A) $(\mu_d - \mu_s) < 0$;
(B) $(\mu_d - \mu_s) > 0$;
(C) $(\mu_d - \mu_s) = 0$.

These are the three physical cases of interest. In case (A), the chemical potential of the solid phase is larger than the chemical potential of the dissolved phase, so it is expected that dissolution will occur spontaneously as dG is negative. This illustrates how a larger chemical potential can be thought of as the tendency of a system to give particles to another. case (B) presents the opposite, where crystallization of the dissolved phase to the solid phase is expected. Here, dG is positive, so the reaction will not occur spontaneously. Finally, case (C) shows chemical equilibrium where no reaction will take place. The first two cases illustrate how a conventional single crystal hydrothermal growth is achieved — shown in Figure 6.5 [37].

In this reaction scheme, a thermal gradient is applied to the different regions so that $dG < 0$ for the two desired cases. In the growth region, a temperature is applied so that $\mu_d < \mu_s$, leading to dissolution — solid phase giving particles to the dissolved

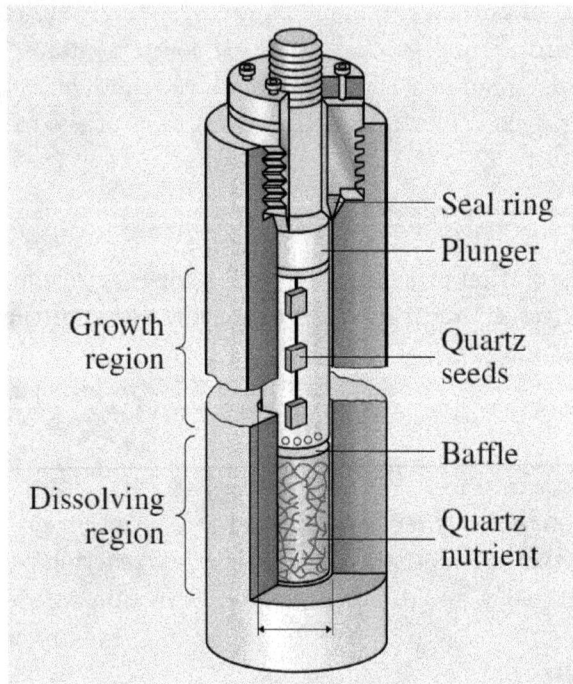

Figure 6.5.   Conventional flat seal hydrothermal appartus whereby a growth region and dissolving region are separated as to enable different thermodynamic mechanisms to drive the chemical reactions in each region from Ref. 36.

phase — while the growth zone is prepared at a temperature so that $\mu_d > \mu_s$, leading to crystallization — dissolved phase giving particles to the solid phase. This is an idealized case, but it describes how the growth of many large single crystals are done even at the industrial level [38–41].

This type of single crystal growth is well-studied and has achieved tremendous results. However, it is idealized and requires the assumption that the dG does not depend on the chemical activity or concentration of other substances in solution. The method of single crystal growth from a seed is not necessarily needed for hydrothermal reactions, especially those done at high temperature and pressures where solvents and solutes can behave drastically different than at standard temperatures and pressure [42, 43].

6.5. Phase Diagrams for Aqueous Solutions

The use of phase diagrams in the traditional flux syntheses of quantum materials is invaluable in assuring phase purity of the desired product. Binary, ternary and even quaternary phase diagrams of elements are catalogued extensively [44–46]. Phase diagrams are essential to remove the empirical trial and error approach to material synthesis, which is still typically required for hydrothermal growth. In general, the Gibb's phase rule determines the number of degrees of freedom which can be manipulated for the system:

$$F = C - P + 2, \tag{6.10}$$

where F is the number of degrees of freedom, C is the number of components in the hydrothermal growth, and P is the number of phases on the phase diagram. The factor of 2 comes from the control of temperature and pressure in typical hydrothermal setup; this can be changed if additional external variables are controlled. The resultant F determines the number of independent intensive variables that can be changed within the system without changing the phase. It is important to note that the number of phases P considered is calculated at a specific temperature and pressure. For example, at the triple point of water, the number of phases is 3, yielding $F = 0$, i.e., the definition of the triple point, a specific temperature and pressure where all three phases co-exist; the conditions cannot be changed without a resulting phase change.

Hydrothermal phase diagrams which encapsulate pressure, temperature and concentration of components of the mixture are sometimes reported in literature but are rare [47–53]. More frequently, diagrams are reported regarding component concentrations at hydrothermal conditions. These are called NC diagrams, where N is the composition of the nutrient and C is the composition for the mineralizing solution (Figure 6.6). These diagrams are used extensively in Russian hydrothermal literature and are well catalogued [54–56]. These are particularly useful in more simple systems but can be used to grow single crystals of particular phases as the crystallization conditions at specific component concentrations are directly listed.

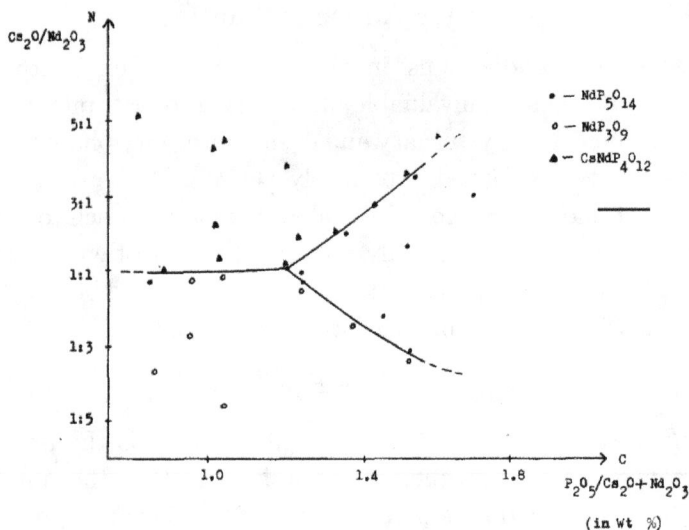

Figure 6.6.   NC diagram for the $Cs_2O-Nd_2O_3-P_2O_5-H_2O$ system from Ref. 56.

From an exploratory synthetic viewpoint, another type of phase diagram is highly instructive when considering what species will exist in solution and thus be able to react to form products. These diagrams are called Pourbaix diagrams. They describe the species present in aqueous solution as a function of electrochemical potential and pH, two variables which can be easily controlled. Pourbaix diagrams for any system can be easily calculated from electrochemical potentials for all species of an element using the standard reduction potential and the van't Hoff and Nernst equations. An example Pourbaix diagram for the speciation of Fe in H_2O is shown in Figure 6.7 [57]. In these diagrams, vertical lines separate phases with no electron transfer, horizontal lines separate phases with no proton transfer, and sloped lines indicate both electron and proton transfer with their slopes proportional to those transfers. These diagrams are often reported at standard temperature and pressure, but additional works have been reported at elevated temperature and pressure akin to hydrothermal conditions [58–61]. They offer a starting point for which exploratory hydrothermal reactions can be designed from first principles.

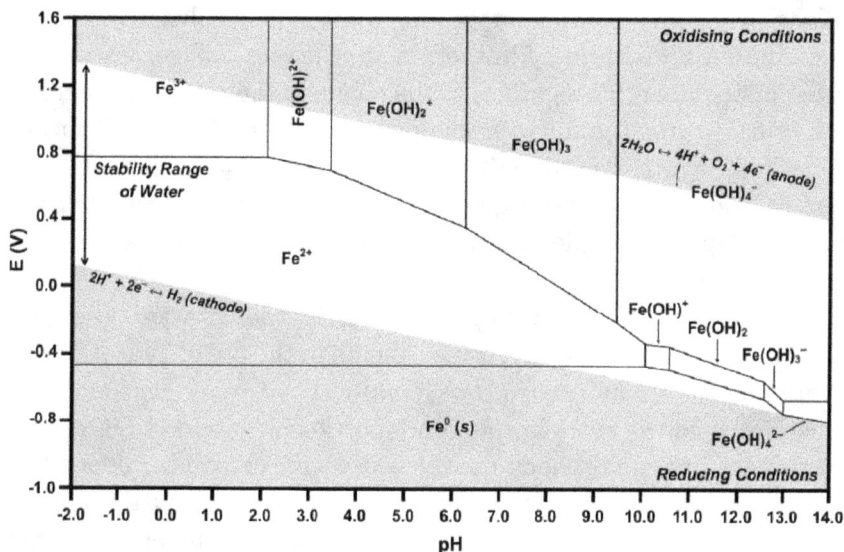

Figure 6.7. Calculated Pourbaix diagram for Fe in H_2O at standard temperature and pressure. The shaded regions shown the stability of water with regards to oxidation and reduction of the solvent from Ref. 57.

Finally, a new approach to hydrothermal synthesis has been rapidly developing over the past 30 years. With the increased computation power of modern systems, thermodynamic modeling of chemical systems at hydrothermal conditions has been explored to intelligently design materials. These methods are capable of modeling very complex aqueous and non-aqueous systems at controllable temperatures, pressures and concentrations. They have been successfully employed in industry and academia to predict and control optimal growth conditions for a wide range of systems [29–31]. This field is growing rapidly and may be key in modernizing hydrothermal synthesis to align with the guided design of quantum materials for the next generation of functional materials.

6.6. Kinetics in Hydrothermal Synthesis

In all different methods of crystal growth, the kinetics of the growth are one of the most important but often least understood aspects. In hydrothermal synthesis, the kinetics is directly controlled by solvation — the interaction of the solutes and solvent. The key confounding variable in hydrothermal synthesis is that solvent–solute interactions are often significant which causes the real hydrothermal solutions to greatly differ from ideal behavior. A central theory is lacking which can describe solubility of all components in a real solution, but guided experimentation can be done through the understanding of simple solvents–solute interactions from a chemical reaction point of view.

The quintessential example of this is the growth of quartz crystals, one of the crowning achievements in the use of hydrothermal synthesis. Quartz naturally grows in natural hydrothermal conditions whereby dissolved silica in the form of orthosilicic acid H_4SiO_4 polymerize to form amorphous quartz which can crystallize depending on the temperature and pressure. Preliminary work on the growth of quartz from laboratory hydrothermal techniques determined that the solubility of quartz in pure water even at high pressures and temperatures was too low for growth to take place over a manageable timeline. It was not until the work of Spezia which showed that quartz solubility in aqueous solutions could be increased by the addition of additional solutes containing hydroxide or halide anions [62]. This study gave a new direction to the growth of crystals from the hydrothermal method. The later seminal work by Laudise systematically reviewed the solubility of quartz as a function of temperature, concentration and pressure and enabled the growth of incredibly large industrially relevant quartz crystal, example shown in Figure 6.8 [63, 64]. This illustrates how the crystal growth and thus kinetics can be controlled and enhanced by the consideration of solute–solute and solute–solvent interactions.

The direct evaluation of the kinetics of crystal growth from the hydrothermal method is often times very difficult. This is because of two important factors: (1) the temperature and pressure required for most growth conditions require significant engineering controls which obscure the ability to probe the reaction media, and (2) most of the reaction media is the solvent which typically interferes with traditional experimental probes. However, there has been several works which use energy-dispersive X-ray diffraction and even neutron

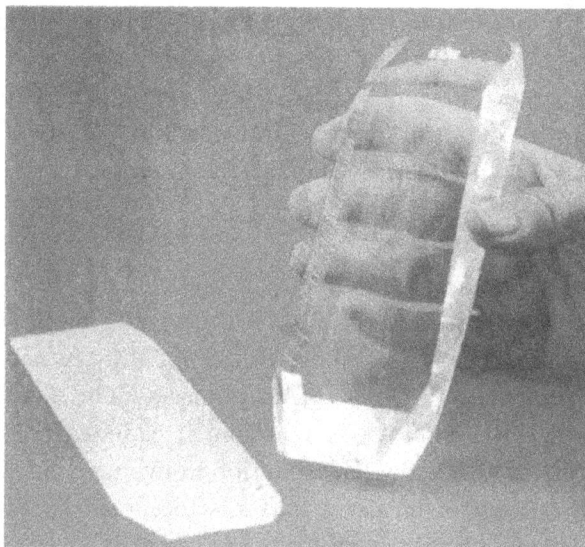

Figure 6.8.   Seed quartz crystal (left) used to grow large single crystals of quartz (right) from the hydrothermal method from Ref. 63.

diffraction to identify the formation of intermediates and products within a hydrothermal reaction [65–70]. These require the use of specifically engineered hydrothermal reaction vessels when high temperature and/or pressure are required, but in the case of mild reaction condition more simple vessels like sealed quartz tubes can be used. For many systems, the evaluation of the kinetics of crystal growth as well as the possible formation of intermediates facilitates the expansion of hydrothermal synthesis of these materials to a more predicative and intelligent design paradigm.

The most frequently used technique for these systems is time-resolved powder diffraction (X-ray or neutron) where the quantitative formation of products and intermediates can be directly evaluated as crystallization occurs. The typical starting point for kinetic analysis of hydrothermal reactions is the Johnson–Mehl–Avrami–Kolmogorov (JMAK) equation which was developed to thermal solid-state transformations [71, 72]:

$$\alpha = 1 - e^{-(k(t-t_0))^n},$$ (6.11)

where α is the phase fraction of product determined by quantitative analysis method, k is the rate constant, t_0 is the crystal growth start time which can be nonzero for many systems and n is the reaction order. Typically, analysis of this data is in the form of a Sharp–Hancock plot which linearizes the JMAK equation to yield:

$$\ln(-\ln(1 - \alpha)) = n \cdot \ln(t - t_0) + n \cdot \ln(k). \qquad (6.12)$$

Sharp and Hancock showed that in the range of $0.15 < \alpha < 0.5$ linear plots will be produced for several crystal growth cases which could be determined by the slope, n, i.e. the JMAK reaction order [73]. Some limitations of the JMAK equation and Sharp–Hancock analysis is that it assumes one reaction mechanism dominates the crystal growth and the JMAK equation convolutes nucleation and growth into one set of terms, n and k, which may be two different processes. To remedy this, several works have attempted to add additional terms to capture more complex nucleation and growth [74–76].

The kinetics of crystal growth from the hydrothermal method is often complex but systematic studies of these kinetics offers avenue for exploitation of hydrothermal growth for targeted synthesis of new quantum materials. Additionally, increased sample environment capabilities in academic laboratories and national user facilities offers unprecedented access into the "black box" of hydrothermal crystal growth.

6.7. The Hydrothermal Apparatus

The necessity for careful consideration of apparatus used for hydrothermal synthesis stems from the high pressure and temperature requirement to obtain hydrothermal conditions. There are many designs, and each have their own uses; the design of particular apparatus for specific systems of study is still required at the edges of academic and industrial research. At the minimum a few requirements must be met for an ideal hydrothermal apparatus or autoclave:

(1) The autoclave must be inert to the desired solvent–solute mixture, i.e., can withstand strong bases, acids, oxidizing agents, and/or reducing agents.

(2) Leak-proof at the desired temperature and pressure range.

(3) Have some sort of built in engineering controls for the mitigation of destructive failure of the vessel — can be external or internal.

(4) Rugged enough or cost-effective enough to be cycled efficiently without requiring additional treatment for every experimental setup.

For the sake of brevity, it is appropriate to only consider types of autoclaves that have been adapted and used worldwide. These are catalogued in Table 6.2 [3, 77, 78].

Most of these listed types of autoclaves are commercially available for use as standalone reaction vessels (Figure 6.2) or within a larger setup such as a hydrothermal furnace (Figure 6.9). Careful consideration must also be taken to ensure the autoclave itself does not have an interaction with the reaction media under consideration. In many cases, the use of inert liners, which sit inside the autoclave to hold the reacting system, is required as to not irreparably damage the autoclave itself. The most commonly used autoclave liner at low temperature, 300°C limit, is Teflon as it is inert in alkaline media unlike Pyrex and quartz. When higher temperatures are needed, liners made from platinum, gold and silver are often required. Platinum and gold offer an upper limit of 700°C with gold typically used for hydroxides and sulphates while platinum and silver are often used for chlorides.

Table 6.2. A short list of the most widely used types of autoclaves for hydrothermal synthesis and research paired with their maximum temperature and pressure limits.

Type of autoclave	Maximum temperature and pressure
Pyrex tube (5mm ID, 9mm OD)	6 bar at 250°C
Quartz tube (5mm ID, 9mm OD)	6 bar at 300°C
Morey type autoclave (flat plate seal)	400 bar at 400°C
Walker and Buehler autoclave (welded)	2000 bar at 480°C
Full Bridgman autoclave	3700 bar at 750°C
Tuttle autoclave (cold cone seal)	5000 bar at 750°C
Opposed anvil autoclave	200 kbar, >1500°C
Opposed diamond anvil autoclave	500 kbar, >1500°C

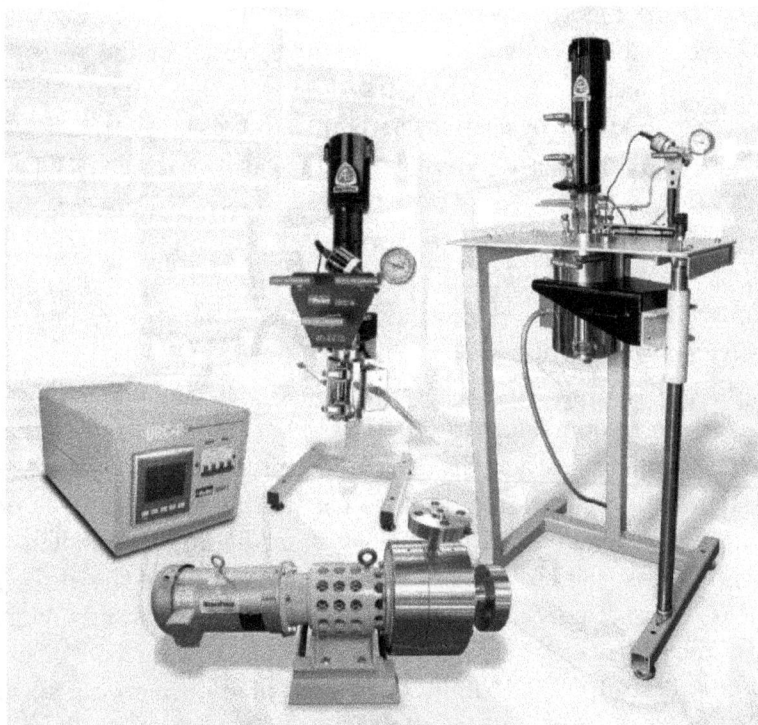

Figure 6.9.   Advertisement photograph for Parker Autoclave Engineers of various models of high pressure and temperature reaction vessels which can be used for hydrothermal reactions. From Parker Autoclave Engineers.

Most commercially available autoclaves require a furnace for temperature control and monitoring. Typically for these setups, the entire autoclave is placed in the furnace. There are more sophisticated setups available where internal heating and/or stirring can be employed and monitored. With all these considerations, the most important aspect of hydrothermal autoclave design and choice is that of safety. High temperatures and pressures required for hydrothermal growth always have the possibility of catastrophic failure from misuse and/or lack of maintenance. In all cases where hydrothermal autoclaves are employed, additional engineering controls should be in place to limit the risk of harm from autoclave failure and volatile chemical evolution from the system.

6.8. Hydrothermal Synthesis of Quantum Materials

Hydrothermal methods are extremely powerful for the growth of quantum materials, especially for halides, oxides and hydroxides. For many covalent oxides, such as quartz, molybdates, titanates, etc., the crystal growth usually requires high temperatures ($>350°C$), whereas the growth of more ionic halides and hydroxides usually takes place below $250°C$. This is explained by the drastic decrease of water dielectric constant above $300°C$, where water behaves as a covalent solvent. For the mild hydrothermal conditions ($<300°C$), it has proven to be invaluable for the synthesis or even the design of quantum materials. Its greatest strength is that the products can be rationally targeted by referring to the Pourbaix diagram and adjusting the pH accordingly. Here, we will discuss spin-liquid materials and Fe-based superconductors, as they represent two distinct hydrothermal approaches for halide/hydroxide and chalcogenide, respectively. The strategies used for their synthesis are translational to materials with similar characteristics. For example, similar conditions used for growing crystals of $ZnCu_3(OH)_6Cl_2$, [79] a quantum spin-liquid can be applied to $LiFePO_4$, a ferrotoroidic material [80].

Quantum spin liquids represent a class of materials show geometric frustration in magnetism [81]. They are relevant for understanding some topological properties and the high-temperature cuprate superconductors. The most prominent examples of quantum spin liquid are materials with kagome lattice such as $ZnCu_3(OH)_6Cl_2$ (Figure 6.10), consisting of corner-sharing $M(OH)_4$ triangles. Freedman *et al.* [79] prepared it by reacting stoichiometric amount of CuO and $ZnCl_2$ hydrothermally at $185°C$ in a quartz tube. For pure crystalline powders, this show no difference from regular Teflon-lined stainless-steel autoclaves. However, the use of quartz ampoule allows further crystal growth utilizing a temperature gradient in a horizontal three-zone furnace. One end of the quartz tube containing $ZnCu_3(OH)_6Cl_2$ is placed in the hot zone of the furnace, and the soluble content is transported to the cold end where slow crystallization is allowed. This temperature gradient hydrothermal method promotes the crystal

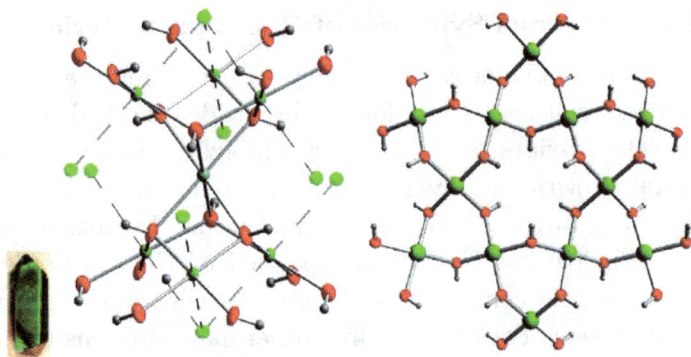

Figure 6.10.   Crystal structure of the spin-liquid $ZnCu_3(OH)_6Cl_2$ with a kagome lattice. The hydrothermally grown green crystals of $ZnCu_3(OH)_6Cl_2$ are shown in the insert. Adapted from [1].

growth up to millimeters, which are large enough to be co-aligned for neutron scattering [79, 82, 83]. The readers should note that the use of quartz ampoules is only possible for mild conditions below 200°C. Hydrothermal reactions above 200°C in a quartz tube can lead to explosion and are extremely dangerous as the vapor pressure of water exceeds 15 atm, close to the pressure limit of regular quartz tubes. For high-temperature crystal growth a more dedicated apparatus using stainless steel autoclave is therefore recommended.

Compared to halides, hydroxides, sulfates, and phosphates, the synthesis of chalcogenides (Ch = S, Se and Te) is less trivial as it requires more careful approach due to possible oxidation or formation of toxic H_2Ch gas. In acidic conditions, large concentrations of HCh^- or Ch^{2-} precursors tend to be oxidized even by proton, as their respectively reduction potentials for S(s) | HS$^-$, Se(s) | HSe$^-$, Te(s) | H$_2$Te are −0.52 V, −0.510 V, −0.595 V, respectively, lower than 0 V for H$_2$ | H$^+$ [84]. Therefore, their synthesis are usually carried out in the highly basic region to avoid oxidization. Such hydrothermal reaction typically work well with more chalcophlic metals such as Zn, Hg, Cd, etc. [3]. Therefore, hydrothermal synthesis of iron-based superconductors was not considered possible until the report of LiOH-intercalated FeSe by Lu *et al.* [85]. To avoid Fe being instantly oxidized to Fe^{3+} in aqueous solution, this synthetic approach uses

Fe powders instead of Fe(II) salts as the Fe source. Fe is charged in an autoclave with selenourea and LiOH to react with water at $120 - 200°C$ for 2–7 days [86–90]. The product $(Li_{1-x}Fe_x)OHFeSe$ ($x = 0.1 - 0.2$) exhibits highest T_c (42–44 K) among the bulk FeSe superconductors. Such structure is stabilized by the partial charge transfer due to Fe doping on the Li site as well as hydrogen bonding from the LiOH layer to the Se atoms in the FeSe layers (Figure 6.11). The Fe substitution in the insulating hydroxide layer not only plays a crucial role to the enhancement of the T_c, but also can induce exotic physical phenomena such as coexistence of magnetic order and superconductivity [87–89].

Similar hydrothermal route can be used to synthesize superconducting FeS ($T_c = 5$ K), analog to the FeSe superconductor, by reaction of Fe powders with Na_2S in water between $100°C$ and $140°C$ for 3–7 days. [91] Interestingly, FeS prepared other than this hydrothermal approach does not show any superconductivity but ferromagnetic behavior [92]. Borg et al. [93] demonstrated that the origin of this magnetism arises from the ferromagnetic thiospinel, Fe_3S_4, impurities using neutron powder diffraction. These results show that this basic hydrothermal approach can preserve the oxidation state of Fe below +2 and lead to superconducting samples. The sulfide analog of $(Li_{1-x}Fe_x)OHFeSe$ can also be prepared using almost identical hydrothermal route, but it was initially reported to be nonsuperconducting [94, 95]. Later, Zhou et al. [96] discovered

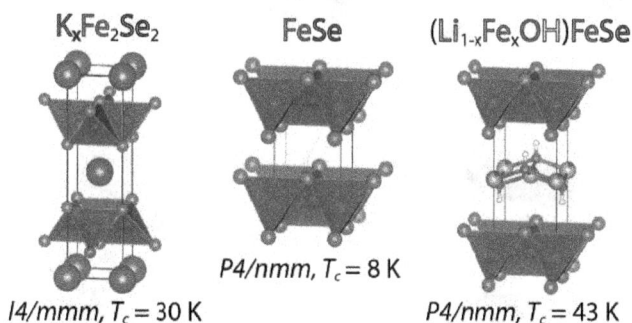

Figure 6.11. Crystal structure of three layered iron selenides: $K_{0.8}Fe_{1.6}Se_2$ (left) FeSe (middle) and $(Li_{1-x}Fe_xOH)FeSe$ (right). Adapted from [90].

that its superconductivity can be improved to 8 K by fine tuning the Fe doping level with more Sn powders in the same autoclave. It is likely that the more reducing Sn kinetically slows the oxidation of Fe, thus allowing more Fe to remain in the LiOH layer after same period of reaction. More interestingly, FeS turns out to be a versatile host subject to intercalation by a variety of adducts hydrothermally (Figure 6.12). Using NaOH instead of LiOH at 160°C, a new NaOH-intercalated FeS can be prepared [96]. This layered sulfide/hydroxide with the composition of $[(Na_{1-x}Fe_x)(OH)_2]FeS$ consists of alternately stacked brucite-type $[(Na_{1-x}Fe_x)(OH)_2]$ and anti-PbO type FeS layers (Figure 6.12). This is quite remarkable compared to the LiOH-intercalated FeS as it is truly a layered heterostructure whereas the former contains two compatible square lattices form for an overall tetragonal structure. When using KOH instead of NaOH, the $K_xFe_{2-y}S_2$ is afforded at $160 - 180°C$, whose synthesis usually requires a much higher temperature above 950°C [96].

Although direct hydrothermal reactions are powerful for synthesizing novel layered structures relevant to the Fe-based superconductors, the as-recovered products contain very tiny

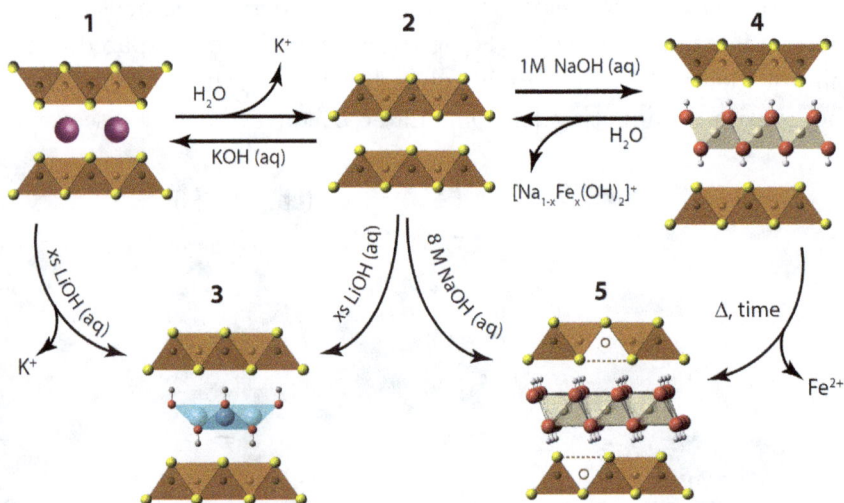

Figure 6.12. Synthetic scheme for the intercalation chemistry of FeS with metal hydroxides and K+ cations via hydrothermal preparations. Adapted from [96].

micron-sized crystals. In order to study their transport properties and magnetism using neutron diffraction, larger sized single crystals are required. Since all the FeCh-based superconductors share the similar anti-PbO-type FeCh layers, crystal conversion is possible by a template method. Because K–Fe–Ch can form a congruent melt, large $K_x Fe_{2-y} Ch_2$ single crystal can be obtained by direct mixing of the respective element and subsequent heating to $1000°C$ followed by slow cool below $500°C$ at $6°C$/hr.

Using the single crystal precursor, centimeter sized FeS single crystals can be obtained by topotactic deintercalation of K from $K_x Fe_{2-y} S_2$ under mild basic hydrothermal conditions $(120 - 160°C)$ [93]. In order to obtain superconducting products, additional Fe powders are required for this reaction as they replenish the Fe vacancy in the template crystals. Similar scheme is also used to convert $K_x Fe_{2-y} Ch_2$ to $(Li_{1-x} Fe_x)OHFeCh$ with excess amount of LiOH in the autoclave other than NaOH or KOH [90, 96]. This kinetically controlled post-synthetic hydrothermal modification can be a generic method to convert a template crystal of thermodynamic stable phase to a metastable phase, which otherwise would not be accessible via traditional solid-state method.

6.9. *In situ* Hydrothermal Studies

Now we have discussed hydrothermal method as a powerful technique for synthesizing and crystal growth of quantum materials. However, because the nature of the "black box" setup, how the reactions progress during hydrothermal reactions are largely unknown. *Ex situ* powder X-ray diffraction at timed intervals can monitor the reaction extent, but it does not shed light on the intermediate phases that are only stable in the hydrothermal environment. Therefore, the most effective technique to study the kinetics and mechanism for hydrothermal reactions is *in situ* X-ray diffraction.

Due to high absorption to low energy X-rays, *in situ* studies are usually carried out using high energy X-ray sources from synchrotron facilities. O'Hare *et al.* pioneered *in situ* diffractions of hydrothermal reactions with large reaction vessels comparable to regular 10–40 mL

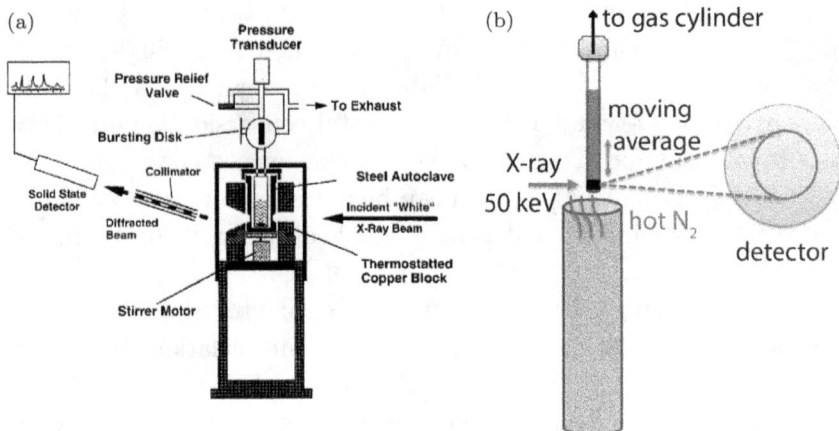

Figure 6.13.   *In situ* diffraction setup for (a) white X-ray source and (b) constant wavelength X-ray source at 50 keV at 17-BM at the advanced photon source. The figure shown in (a) is adapted from [97].

sized autoclaves [97–100]. The instrumental setup using white X-ray spectra between 5 and 140 keV at the UK Synchrotron Radiation Source at Daresbury Laboratory is shown in Figure 6.13(a). [97]. This setup stainless steel autoclave with part the stainless steel wall machined down to 0.7 mm to allow X-ray beam to pass through. The detector is fixed at the angle θ, and the Bragg reflection condition is met with $E = n/d\sin\theta$, where E is energy, n is the allows the use of a regular sized Parr-Instrument geometric constant relevant to the detector distance. Thus, the d-space is as a function of energy E. Although this setup most accurately reflects a lab hydrothermal reaction in size and quantity, it only covers very limited d-space (4–13 Å). Moreover, when the energy is close to the absorption edge of a certain element, the appearance of a huge resonance peak may interfere with the diffraction peaks of interests. Therefore, an easier setup with thin capillary using constant wavelength is more popular among *in situ* diffraction studies.

Many synchrotron beamlines, such as P02.1 at DESY, BM01A at ESRF and BL02B2 at Spring-8, allow the use of 0.7–1.1 mm capillaries as the reaction vessels [100–104]. The capillaries are usually connected to a syringe pump that can accommodate a very high pressure. For example, when using sapphire capillary at BL02B2 at Spring-8, the maximum pressure can reach up to 300 atm, thus capable of heating the capillary above 300°C [105]. Although this setup is quite powerful, the use of a syringe pump with a small capillary is not ideal for reproducing the hydrothermal environment of regular laboratory reaction vessels. For example, the small sample size can be considered negligible compared to the large volume of solvent in the head space of the syringe pump. Therefore, a more versatile setup shown in Figure 6.13(b) is developed with the assistance of Dr. Andrey Yakovenko at the beamline 17-BM at the Advanced Photon Source (APS). This setup uses a regular He or Ar gas instead of liquid to pressurize reaction vessels. For low pressure reactions, it can be directly connected to a regular gas cylinder capable of reaching 30 atm or 220°C. In addition, the size of the vessel can be as large as 3.1 mm which allows more materials. In addition, larger capillaries can avoid trapped bubbles due to low surface tension compared to smaller ones. The heater is placed beneath the sample which allows more uniform temperature compared to horizontally placed heaters at other beamlines. The whole transparent setup allows real-time monitoring of the sample location. Moreover, during reactions a motor can oscillate the capillary vertically to collect the average information from the entire sample. This new setup enables rapid screening of *in situ* hydrothermal reactions. The sample change and alignment takes merely 30–60 min.

6.10. Electro-hydrothermal Synthesis

As mentioned in Section 6.9, when the synthetic temperature is below 300°C, the dielectric constant of water is large. Therefore, the soluble ionic species, such as alkali bases or salts, turn the solution into an electrolyte, where all redox reactions can be predicted by the Pourbaix diagram. This can be utilized for phase selection or rational design of novel structures. For example, this electrochemical nature has been exploited for the synthesis of $BaTiO_3$ films as shown in Figure 6.14 [106]. In this ingenious setup, a titanium anode and a platinum cathode are immersed in $Ba(OH)_2$

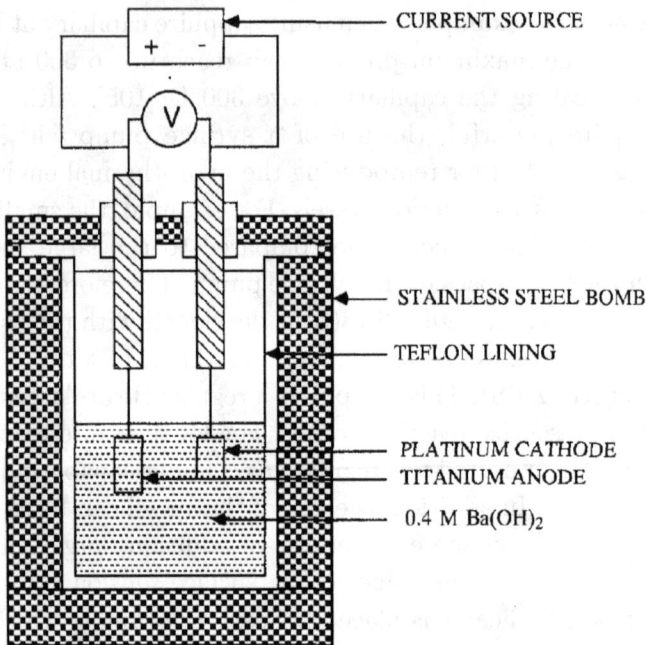

Figure 6.14.   A schematic illustration of the electrochemical-hydrothermal setup for the synthesis of BaTiO$_3$ films. Adapted from [29].

solution inside a stainless steel autoclave. A voltage is applied to the electrodes that are connected to a current source outside the autoclave, thus composing an electrolysis cell. Titanium metal exhibits high resistance to corrosion, but by applying a large voltage in hydrothermal conditions, the electrochemical driving force is high enough to proceed this reaction quickly, which takes barely 30–120 min to complete at around 100–200°C. This is an extreme example that how the electrolytic nature of the solution can be utilized for electrochemical synthesis during hydrothermal reactions. As a matter of fact, at elevated temperatures the in situ electrochemical driving force is high enough for spontaneous reactions to proceed.

Let us look at the reaction of $(Li_{1-x}Fe_x)OHFeSe$ again. The critical setup for this reaction is the digestion of Fe metal. Of course, electrochemical corrosion is commonly seen for Fe as rust (iron oxides) is a part of everyday life. However, we have to emphasis that the hydrothermal reaction of $(Li_{1-x}Fe_x)OHFeSe$ is under an oxygen-free environment. Thus, the only oxidizing source is water where Fe reacts with water to form $Fe(OH)_2$ and H_2. It is very surprising that this reaction can occur at barely above $100°C$, in contrast to the same reaction with water vapor above $500°C$. However, this is quite reasonable if we take the electrochemical driving force into account. The saturated LiOH solution acts as an electrolyte, Fe as anode, water as cathode. Thus, we can write this cell reaction as $Fe(aq)|\ Fe(OH)_2(s)\|\ LiOH(aq)\|\ H_2O(l)|\ H_2(g)$, where the anode and cathode potentials are -0.89 V and -0.828 V, respectively [19]. Therefore, there is a small yet positive (0.062 V) driving force for Fe oxidation. With this knowledge, the reaction mechanism for the hydrothermal synthesis of $(Li_{1-x}Fe_x)OHFeSe$ is more clear. In this reaction, Fe metal is slowly oxidized to $Fe(OH)_2$, which quickly reacts with HSe^- in the solution so that it does not further oxidize to Fe^{3+}. This setup is critical for preserving the superconductivity. Sun *et al.* [89]. report that LiOH can directly intercalate into anti-PbO type FeSe to form $(Li_{1-x}Fe_x)OHFeSe$ under the same hydrothermal condition, but the products are never superconducting unless about 6% of Fe powders are added to the autoclave. Similar reducing effect of Fe metal is observed for the conversion of FeS single crystals using $K_xFe_{2-y}S_2$ [93].

Understanding this electrochemical nature of this basic hydrothermal reaction is the key to rational design. Now we have a series of metals are subject to this digestion process such as Mg $(-2.70$ V), Sn $(-1.839$ V), Mn $(-1.565$ V), Cr $(-1.48$ V), Zn $(-1.249$ V), Mo $(-0.98$ V) etc. As long as the reduction potential is below that of water $(-0.828$ V), there is an electrochemical driving force for the reaction to proceed. The advantage of this slow metal digestion approach is that it allows molecular species to form in solution and polymerize into template sheets to co-precipitate as heterolayered structures. In contrast, reacting soluble salts such as $FeCl_2$ with Na_2S in aqueous solution only results in fast precipitation of an amorphous phase instead of layered FeS [92]. Therefore, this slow electrochemical digestion can be used for designing novel heterolayered materials.

Let us take a look at $[(Na_{1-x}Fe_x)(OH)_2]FeS$ mentioned in the previous section. When using lower NaOH concentration (1 M), Fe reacts with excess thiourea to from binary FeS [96]. If 5 M NaOH is used with Fe/thiourea ration >1, this heterolayered structure is formed. Excess $Fe(OH)_2$ is oxidized to +3 in the brucite type $[(Na_{1-x}Fe_x)(OH)_2]^{\delta+}$ layer without enough sulfur source and it co-precipitate with $FeS^{\delta-}$ layer to form $[(Na_{1-x}Fe_x)(OH)_2]FeS$. This is only possible with this *in situ* slow digestion process. It is found that directly reacting excess NaOH with FeS only result in conversion of FeS to more thermodynamic stable Fe_3O_4. Therefore, it is possible to use other metals such as Mg, Mn and Zn to replace Fe in the brucite layer.

Here, we formulate a simple geometrical argument to explain how to design such heterostructures. To construct the smallest regular unit cell, we inscribe a tetragonal one and hexagonal one within a larger square cell (Figure 6.15(a)). The larger square cell is a $\sqrt{2} \times \sqrt{2}$ superstructure of the tetragonal one with lattice constant a_T. If the hexagonal cell's lattice parameter is a_H, then we can define a tolerance factor τ as being the simple ratio a_T/a_H. To form an undistorted heterostructure, τ should be $\sqrt{3}/\sqrt{2} = 1.22$. Therefore, the closer τ is to 1.22, the more compatible the hexagonal and tetragonal lattices are, and larger or smaller τ should result in stress or strain to the FeSe layer, respectively. Thus, we can use this to rationally design heterolayered structure exhibit both 4-fold and 6-fold symmetries similar to $[(Na_{1-x}Fe_x)(OH)_2]FeS$.

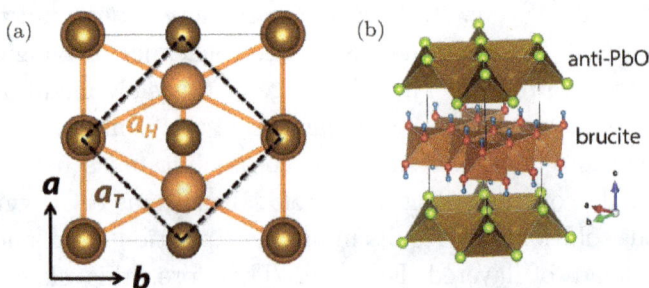

Figure 6.15.   The smallest unit cell (a) of a heterolayered structure consisting of both 4-fold and 6-fold symmetry and the proposed structure (b).

We have shown that electrochemical potentials can be used to our advantage for the synthesis of novel heterolayered materials, which is dependent on the electrolytes of alkali base solutions. However, alkali bases tend to intercalate themselves as demonstrated by interaction chemistry of FeS with LiOH, NaOH and KOH [96]. When their intercalation is not desired, we need to an alternative strong base to from an electrolyte and digest metals. Zhou *et al.* [107] report the synthesis of ammonia-intercalated FeCh (NH$_{3.5}$FeCh) under hydrothermal conditions using an organic metal-free base– guanidine, $(NH_2)_2C=NH$. Guanidine, analogous to urea, exhibits a basicity (pKa = 13.6) comparable to KOH in aqueous solution. Therefore, it can completely dissociate to OH$^-$ and $(NH_2)_2C=NH_2^+$ in aqueous solution to form an electrolyte and effectively digest Fe metal powders under hydrothermal conditions. This awesome approach using organic bases such as guanidine to replace alkali bases can allow species with weaker intercalation potentials, such as ammonia, to form heterolayered compounds with host layers such as FeS.

The methods describes in this section lay groundwork for rational design of novel layered quantum materials via electrochemical potential under hydrothermal conditions. Finally, we want to point out that use of heavy water, D_2O, instead of water, H_2O may enable selection of thermodynamic products. Although, isotope effect has not been reported for product selection, it is quite possible with highly basic hydrothermal conditions. Zhou *et al.* discovered that when D_2O is used for the hydrothermal synthesis of the $(Li_{1-x}Fe_x)ODFeSe$ superconductors, iron oxide impurities are much less likely to form compared to their H_2O counterparts [90]. Such impurities are extremely easy to form for reactions above 140°C for longer than 3 days unless Sn metals are added to the autoclave to create a more reducing environment. This hints for lower reduction potential of D_2O compared to H_2O under highly basic conditions. This is indeed the case how deuterium was discovered — electrolysis of saturated alkali solution by Harold Urey *et al.* [108]. Thus the reader should also take this isotope effect into consideration when using hydrothermal methods.

Acknowledgments

The author would like to thank Dr. Andrey Yakovenko at the beamline 17-BM at the Advanced Photon Source (APS) for the assistance to develop the *in situ* hydrothermal setup. This research used resources of the Advanced Photon Source, a US Department of Energy (DOE) Office of Science User Facility operated for the DOE Office of Science by Argonne National Laboratory under Contract No. DE-AC02-06CH11357.

Bibliography

[1] V. G. Tarasov, A. V. Gebruk, A. N. Mironov, and L. I., Moskalev, *Chem. Geol.*, **224**, 5 (2005).
[2] D. L. Norton, *Annu. Rev. Earth Planet. Sci.*, **12**, 155 (1984).
[3] K. Byrappa, and M. Yoshimura, *Handbook of Hydrothermal Technology* (Elsevier Science, 15 2012).
[4] S. Feng, and X. Ruren, *Acc. Chem. Res.*, **34**, 239 (2001).
[5] M. S. Whittingham, *Curr. Opin. Solid St. M.*, **1**, 227 (1996).
[6] T. Chirayil, P. Zavalij, and M. S. Whittingham, *Chem. Mater.*, **10**, 2629 (1998).
[7] C. S. Cundy, and P. A. Cox, *Chem. Rev.*, **103**, 663 (2003).
[8] C. S. Cundy, and P. A. Cox, *Micropor. Mesopor. Mat.*, **82**, 1 (2005).
[9] X.-M. Chen, and M.-L. Tong, *Acc. Chem. Res.*, **40**, 162 (2007).
[10] K. Byrappa, and T. Adschiri, *Prog. Cryst. Growth Ch.*, **53**, 117 (2007).
[11] W. Han, Y. Qiu, Y. Zhao, H. Zhang, J. Chen, S. Sun, L. Lan, Q. Fan, and Q. Li, *Cryst. Eng. Comm.*, **18**, 7934 (2016).
[12] N. L. Rosi, J. Eckert, M. Eddaoudi, D. T. Vodak, J. Kim, M. O'Keefe, and O. M. Yaghi, *Science*, **300**, 1127 (2003).
[13] R. M. Barrer, *Chem. Brit.*, **2**, 380 (1966).
[14] Image of Parr autoclaves sold by Latech Scientific Supply Pte. Ltd., Latech.com.sg.
[15] A. Rabenau, *Angew. Chem.*, **24**, 1026 (1985).
[16] R. I. Walton, *Chem. Soc. Rev.*, **31**, 230 (2002).
[17] D. R. Modeshia, and R. I. Walton, *Chem. Soc. Rev.*, **39**, 4303 (2010).
[18] G. Demazeu, *J. Mater. Sci.*, **43**, 2104 (2008).
[19] D. R. Lide, *CRC Handbook of Chemistry and Physics*, Vol. 85 (CRC Press, 2004).
[20] H. B. Callen, *Thermodynamics & An Introduction to Thermostatistics* (Wiley, sons, 2006).

[21] D. A. Palmer, R. Fernandez-Prini, and A. H. Harvey, *Aqueous Systems at Elevated Temperatures and Pressures* (Elsevier, 2004).

[22] P. R. Tremaine, and N. R. C. Canada, Ontario Power Generation, I. *Steam, Water, and Hydrothermal Systems: Physics and Chemistry Meeting the Needs of Industry: Proceedings of the 13th International Conference on the Properties of Water and Steam.* (NRC Research Press, 2000).

[23] V. Valyashko, *Hydrothermal Properties of Materials: Experimental Data on Aqueous Phase Equilibria and Solution Properties at Elevated Temperatures and Pressures* (Wiley, 2008).

[24] G. Brunner, *Hydrothermal and Supercritical Water Processes*; (Elsevier Science, 2014).

[25] M. M. Lencka, and R. E. Riman, *Chem. Mater.*, **5**, 61 (1993).

[26] M. M. Lencka, and R. E. Riman, *J. Am. Ceram. Soc.*, **76**, 2649 (1993).

[27] S. Venigalla, and J. H. Adair, *Chem. Mater.*, **11**, 589 (1999).

[28] A. Testino, V. Buscaglia, M. T. Buscaglia, M. Viviani, and P. Nanni, *Chem. Mater.*, **17**, 5346 (2005).

[29] A. Dias, and V. S. T. Ciminelli, *J. Eur. Ceram. Soc.*, **21**, 2061 (2001).

[30] A. Anderko, S. J. Sanders, and R. D. Young, *Corrosion*, **53**, 43 (1997).

[31] R. E. Riman, W. L. Suchanek, and M. M. Lencka, *Ann. Chim. Scie. Mater.*, **27**, 15 (2002).

[32] J. M. Prausnitz, R. N. Lichtenthaler, and E. G. de Azevedo, *Molecular Thermodynamics of Fluid-Phase Equilibria* (Pearson Education, 1998).

[33] R. A. Robinson, and R. H. Stokes, *Electrolyte Solutions: Second Revised Edition* (Dover Publications, 2002).

[34] J. S. Rowlinson, F. L. Swinton, J. E. Baldwin, A. D. Buckingham, and S. Danishefsky, *Liquids and Liquid Mixtures* (Elsevier Science, 2013).

[35] M. Pekavr, and I. Samohyl, *The Thermodynamics of Linear Fluids and Fluid Mixtures* (Springer, 2014).

[36] E. Kaldis, *Crystal Growth of Electronic Materials* (North-Holland, 1985).

[37] R. A. Laudise, E. D. Kolb, and A. J. Caporaso, *J. Am. Ceram. Soc.*, **47**, 9 (1964).

[38] R. Feigelson and R. Laudise, *50 years progress in crystal growth: A reprint collection.* (Elsevier, 2004).

[39] J. M. Stanley, *Ind. Eng. Chem.*, **46**, 1684 (1954).

[40] A. C. Walker, and E. Buehler, *Ind. Eng. Chem.*, **42**, 1369 (1950).

[41] J. Arends, J. Schuthof, W. H. van der Lindwen, P. Bennema, and P. J. Van Der Berg, *J. Cryst. Growth*, **46**, 213 (1979).

[42] K. Kodaira, Y. Iwase, A. Tsunashima, and T. Matsushita, *J. Cryst. Growth*, **60**, 172 (1982).

[43] M.-Y. Huang, Y.-H. Chen, B.-C. Chang, and K.-H. Lii, *Chem. Mater.*, **17**, 5743 (2005).

[44] H. Okamoto, *ASM Int.*, **44**, (2010).

[45] P. Villars, A. Prince, H. Okamoto, and others. *Handbook of Ternary Alloy Phase Diagrams*, Vol. 5 (ASM International Materials, 1995).

[46] G. B. Stringfellow, *J. Cryst. Growth*, **27**, 21 (1974).

[47] M. Gehrig, H. Lentz, and E. U. Franck, *Berichte der Bunsengesellschaft Phys. Chemie*, **90**, 525 (1986).

[48] H. Weingärtner, and E. U. Franck, *Angew. Chemie Int. Ed.*, **44**, 2672 (2005).

[49] D. Y. Zezin, A. A. Migdisov, and A. E. Williams-Jones, *Geochim. Cosmochim. Acta*, **75**, 5483 (2011).

[50] H. Shinohara, and K. Fujimoto, *Geochim. Cosmochim. Acta*, **58**, 4857 (1994).

[51] I. M. Abdulagatov, and N. D. Azizov, *Int. J. Thermophys.*, **24**, 1581 (2003).

[52] C. Schmidt, and R. J. Bodnar, *Geochim. Cosmochim. Acta*, **64**, 3853 (2000).

[53] R. A. Laudise, W. A. Sunder, R. F. Belt, and G. Gashurov, *J. Cryst. Growth*, **102**, 427 (1990).

[54] O. S. Bondareva, Y. A. Malinovskij, I. P. Kuz'mina, and V. A. Kuznetsov, *Kristallografiya*, **31**, 159 (1986).

[55] B. N. Litvin, V. S. Fonin, D. Y. Pushcharovskii, and E. A. Pobedimskaya, in *Pocm Kpucmannob/Rost Kristallov/Growth of Crystals* (Springer, 1984), pp. 94–98.

[56] B. N. Litvin, and K. Byrappa, *J. Cryst. Growth*, **51**, 470 (1981).

[57] D. Channei, S. Phanichphant, A. Nakaruk, S. S. Mofarah, P. Koshy, and C. C. Sorrell, *Catalysts*, **7**, 45 (2017).

[58] B. Beverskog, and I. Puigdomenech, *Corros. Sci.*, **38**, 2121 (1996).

[59] B. Beverskog, and I. Puigdomenech, *Corros. Sci.*, **39**, 969 (1997).

[60] L. B. Kriksunov, and D. D. Macdonald, *Corrosion*, **53**, 605 (1997).

[61] M. H. Kaye, and W. T. Thompson, *Uhlig's Corros. Handb.*, **3**, 111 (2011).

[62] G. Spezia, *Atti della R. Accademia delle scienze di Torino*, volume 41, page 158.

[63] R. A. Laudise, J. W. Nielsen, F. Seitz, and D. Turnbull, (Eds), *Solid State Phys.* **12**, 149 (1961).

[64] R. A. Laudise, *Chem. Eng. News Arch.*, **65**, 30 (1987).

[65] K. M. Peterson, P. J. Heaney, and J. E. Post, *Chem. Geol.*, **444**, 27 (2016).

[66] J. Cravillon, C. A. Schröder, H. Bux, A. Rothkirch, J. Caro, and M. Wiebcke, *Cryst. Eng. Comm.*, **14**, 492 (2012).

[67] A. M. Fogg, and D. O'Hare, *Chem. Mater.*, **11**, 1771 (1999).

[68] A. M. Beale, and G. Sankar, *Chem. Mater.*, **15**, 146 (2003).

[69] R. I. Walton, F. Millange, R. I. Smith, T. C. Hansen, and D. O'Hare, *J. Am. Chem. Soc.*, **123**, 12547 (2001).

[70] A. Michailovski, R. Kiebach, W. Bensch, J.-D. Grunwaldt, A. Baiker, S. Komarneni, and G. R. Patzke, *Chem. Mater.*, **19**, 185 (2007).

[71] M. Avrami, *J. Chem. Phys.*, **7**, 1103 (1939).

[72] J. William, and R. Mehl, *Trans. Met. Soc. AIME*, **135**, 416 (1939).

[73] J. D. Hancock, and J. H. Sharp, *J. Am. Ceram. Soc.*, **55**, 74 (1972).

[74] B. Rheingans, and E. J. Mittemeijer, *JOM*, **65**, 1145 (2013).

[75] J. Wang, H. C. Kou, X. F. Gu, J. S. Li, L. Q. Xing, R. Hu, and L. Zhou, *Mater. Lett.*, **63**, 1153 (2009).

[76] L. A. Pérez-Maqueda, J. M. Criado, and P. E. Sánchez-Jiménez, *J. Phys. Chem. A*, **110**, 12456 (2006).

[77] K. Byrappa, in Springer Handbook of Crystal Growth, Dhanaraj, G.; Byrappa, K.; Prasad, V.; Dudley, M., (Eds.), (Springer, 2010), pp. 599–653.

[78] P. Rudolph, Handbook of Crystal Growth, (Elsevier Science, 2014).

[79] D. E. Freedman, T. H. Han, A. Prodi, P. Müller, Q.-Z. Huang, Y.-S. Chen, S. M. Webb, Y. S. Lee, T. M. McQueen, and D. G. Nocera, *J. Am. Chem. Soc.* **132**, 16185 (2010).

[80] S. Gnewuch, and E. E. Rodriguez, *J. Solid State Chem.*, **271**, 175 (2019).

[81] P. W. Anderson, *Mater. Res. Bull.*, **8**, 153 (1973).

[82] T.-H. Han, J. S. Helton, S. Chu, D. G. Nocera, J. A. Rodriguez-Rivera, C. Broholm, and Y. S. Lee, *Nature*, **492**, 406 (2012).

[83] T. H. Han, J. S. Helton, S. Chu, A. Prodi, D. K. Singh, C. Mazzoli, P. Muller, D. G. Nocera, and Y. S. Lee, *Phys. Rev. B*, **83**, 100402 (2011).

[84] M. Bouroushian, in *Electrochemistry of Metal Chalcogenides* (Springer, 2010), p 57.

[85] X. F. Lu, N. Z. Wang, G. H. Zhang, X. G. Luo, Z. M. Ma, B. Lei, F. Q. Huang, and X. H. Chen, *Phys. Rev. B*, **89**, 020507 (2014).

[86] U. Pachmayr, F. Nitsche, H. Luetkens, S. Kamusella, F. Brückner, R. Sarkar, H.-H. Klauss, and D. Johrendt, *Angew. Chem. Int. Ed.*, **54**, 293 (2015).

[87] X. F. Lu, N. Z. Wang, H. Wu, Y. P. Wu, D. Zhao, X. Z. Zeng, X. G. Luo, T. Wu, W. Bao, G. H. Zhang, F. Q. Huang, Q. Z. Huang, and X. H. Chen, *Nat. Mater.*, **14**, 325 (2015).

[88] J. W. Lynn, X. Zhou, C. K. H. Borg, S. R. Saha, J. Paglione, and E. E. Rodriguez, *Phys. Rev. B*, **92**, 060510 (2015).

[89] H. Sun, D. N. Woodruff, S. J. Cassidy, G. M. Allcroft, S. J. Sedlmaier, A. L. Thompson, P. A. Bingham, S. D. Forder, S. Cartenet, N. Mary, S. Ramos, F. R. Foronda, B. H. Williams, X. Li, S. J. Blundell, and S. J. Clarke, *Inorg. Chem.*, **54**, 1958 (2015).

[90] X. Zhou, C. K. H. Borg, J. W. Lynn, S. R. Saha, J. Paglione, and E. E. Rodriguez, *J. Mater. Chem. C*, **4**, 3934 (2016).

[91] X. Lai, H. Zhang, Y. Wang, X. Wang, X. Zhang, J. Lin, and F. Huang, *J. Am. Chem. Soc.*, **137**, 10148 (2015).

[92] I. T. Sines, D. D. Vaughn II, R. Misra, E. J. Popczun, and R. E. Schaak, *J. Solid State Chem.*, **196**, 17 (2012).

[93] C. K. H. Borg, X. Zhou, C. Eckberg, D. J. Campbell, S. R. Saha, J. Paglione, and E. E. Rodriguez, *Phys. Rev. B*, **93**, 094522 (2016).

[94] U. Pachmayr, and D. Johrendt, *Chem. Comm.*, **51**, 4689 (2015).

[95] X. Zhang, X. Lai, N. Yi, J. He, H. Chen, H. Zhang, J. Lin, and F. Huang, *RSC Adv.*, **5**, 38248 (2015).

[96] X. Zhou, C. Eckberg, B. Wilfong, S.-C. Liou, H. K. Vivanco, J. Paglione, and E. E. Rodriguez, *Chem. Sci.*, **8**, 3781 (2017).

[97] R. J. Francis, S. J. Price, J. S. O. Evans, S. O'Brien, D. O'Hare, S. M. Clark, *Chem. Mater.*, **8**, 2102 (1996).

[98] R. J. Francis, S. O'Brien, A. M. Fogg, P. S. Halasyamani, D. O'Hare, T. Loiseau, and G. Férey, *J. Am. Chem. Soc.*, **121**, 1002 (1999).

[99] R. J. Francis, and D. O'Hare, *J. Chem. Soc., Dalton Trans.*, **19**, 3133 (1998).

[100] R. I. Walton, F. Millange, D. O'Hare, A. T. Davies, G. Sankar, and C. R. A. Catlow, *J. Phys. Chem. B*, **105**, 83 (2001).

[101] E. D. Bojesen, K. M. Ø. Jensen, C. Tyrsted, N. Lock, M. Christensen, and B. B. Iversen, *Cryst. Growth Des.* **14**, 2803 (2014).

[102] P. Nørby, M. Roelsgaard, M. Søndergaard, and B. B. Iversen, *Cryst. Growth Des.*, **16**, 834 (2016).

[103] N. Lock, E. M. L. Jensen, J. Mi, A. Mamakhel, K. Norén, M. Qingbo, and B. B. Iversen, *J. Chem. Soc. Dalton Trans.*, **42**, 9555 (2013).

[104] Y. Sun, and Y. Ren, *Part. Part. Sys. Char.*, **30**, 399 (2013).

[105] T. Fujita, H. Kasai, and E. Nishibori, *J. Supercrit. Fluids* **147**, 172 (2019).

[106] R. Bacsa, P. Ravindranathan, and J. Dougherty, *J. Mater. Res.*, **7**, 423 (1992).

[107] X. Zhou, B. Wilfong, S.-C. Liou, H. Hodovanets, C. M. Brown, and E. E. Rodriguez, *Chem. Comm.*, **54**, 6895 (2018).

[108] H. C. Urey, F. G. Brickwedde, and G. M. Murphy, *Phys. Rev.*, **39**, 164 (1932).

Chapter 7

Floating Zone Crystal Growth

Eli B. Zoghlin* and Stephen D. Wilson[†]

Materials Department, University of California,
Santa Barbara CA USA 93106, USA
*eaz@ucsb.edu
[†]stephendwilson@ucsb.edu

7.1. Introduction and Overview

The floating zone crystal growth technique arose from the contemporaneous development by Pfann and Theurer (at Bell Telephone Laboratories) [1] and by Keck *et al.* (at the Army Signal Corps Laboratories) [2, 3] of a remote means of melting, purifying, and recrystallizing semiconducting material. Either radiant or inductive heating was employed to establish a focused heating zone that locally melts a compound whose melt would then remain suspended via surface tension between solid regions of the material outside of the heating zone. Recrystallization then proceeds as the molten material passes out of the heated zone, and the approach was celebrated as a crucible-free or "floating" means of crystal growth and zone refining purification. The technique was pivotal in the early 1950s for the initial production of semiconductor-grade Ge and Si with impurity concentrations below one part per billion needed for the nascent transistor industry, and it remains at the forefront of the growth of ultrahigh purity electronic and quantum materials today.

This chapter focuses on modern incarnations of floating zone crystal growth, which now predominantly revolve around the use of optical heating sources, for the growth of quantum materials in fundamental research. As such, we do not attempt to cover industrial/applied applications of the technique used, for instance, in the semiconductor and scintillator industries. Instead, our discussion focuses on the approaches, methodologies, and applications of the floating zone technique for the exploration of new quantum materials, many of which host complex energy landscapes strongly renormalized by extrinsic disorder. Due to this sensitivity to disorder, ultrahigh purity single crystal specimens of these materials are essential for experimentally realizing the delicate phases predicted to arise from the extended and competing interactions in their underlying ground states. The technique remains pivotal to advances in fields as diverse as high temperature superconductivity, frustrated magnetism, strongly correlated electron physics, and increasingly in the stabilization of topologically nontrivial electronic states.

In the following sections, we first describe the principle of the floating zone technique as well as highlight examples of its use in the growth of single crystals of quantum materials. Variations of the technique for incongruently melting and volatile materials such as the traveling solvent floating zone technique are then described, followed by a description of the types and operation of typical research-grade floating zone furnaces. We conclude with an overview of floating zone crystal growth at environmental extremes such as the rapidly developing field of high-pressure floating zone growth and briefly discuss the near-term outlook for the technique. We also direct the interested reader to several other recent reviews of the technique and its applications [4–6].

7.2. The Floating Zone Growth Method

7.2.1. *Description of the technique and principles*

Incorporating the idea of zone-refining, the floating zone (FZ) technique entails growing a single crystal from a crucible-free melt, which is passed through feed material of the appropriate stoichiometry. The general procedure for such a growth is as follows. Polycrystalline rods of the compound to be grown are secured above and below (the feed and seed rods, respectively) a small, well-defined hot zone. The heating power is then slowly increased until the material in the hot zone is completely molten and the two rods can be joined via a liquid "bridge". This floating, molten zone between the two rods is maintained by surface tension. A sample is then grown by continuously feeding polycrystalline material into the molten zone, which subsequently recrystallizes as it exits. This is done by either moving the heating source upwards along the rods or by translating the rods in tandem through the stationary hot zone. Smooth translation and rotation of the feed and seed rods is obtained by attaching them to the ends of drive shafts which pass into the growth chamber.

Upon exiting the hot zone and cooling, multiple grains initially nucleate from the polycrystalline seed and grow simultaneously from the melt. Grain selection occurs as grains with faster radial growth velocities outcompete others, until a single crystal grain is obtained. This process is illustrated generically in Figure 7.1, with a schematic of a typical FZ furnace for reference. Once a crystal has been produced, it can then serve as a seeding template for future crystal growths and replace the polycrystalline seed — some materials require several reseeded growths before a single crystalline boule is obtained.

In typical operation, the molten zone is maintained until the feed rod has been consumed, at which point translation is then stopped

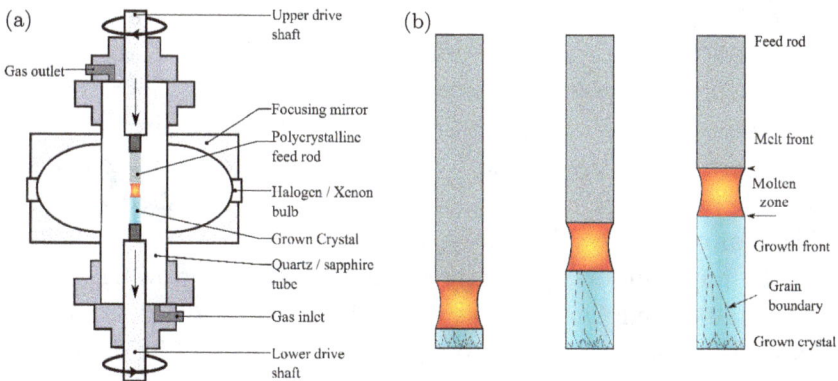

Figure 7.1. (a) Schematic of an optical floating zone setup. (b) Diagram of the floating zone growth process showing the formation of a single crystal via grain growth and subsequent grain selection (left to right). The grain which eventually dominates in the rightmost image possesses the highest radial growth velocity.

and the feed rod is slowly moved out of the hot zone and separated from the grown crystal. The tip of the crystal is kept within the hot zone during separation and as the heating power is ramped down, to minimize thermal shock. In exploratory studies, the molten zone may alternatively be quenched if its composition is of interest. All floating zone furnaces are designed to maintain a stable molten zone for continuous growth over a long period of time and as such there are several universal features. A finely focused hot zone, of a consistent temperature, with a sharp axial heating gradient is contained within an optically accessible chamber supporting a gas environment. This provides control of the growth atmosphere, pressure and flow conditions. Optical access is required for power delivery in lamp and laser-based methods (though not necessarily for induction heating); however, optical access is essential for observing and tuning the growth process *in situ*, rendering FZ-growth an "active" crystal growth technique, similar to the Czochralski method discussed in Chapter 5 [7]. In contrast, crucible-based flux (Chapters 3 and 7) or vapor transport (Chapter 4) methods are "blind" techniques. Here, the success of a growth run must be determined after its completion and the results used iteratively to inform future modifications of the growth parameters.

Before discussing the variety of growth parameters available in FZ-growth, it is beneficial to review the two advantages of this technique, with respect to crystal purity, which are especially important for quantum materials synthesis. The first of these is the crucible-free growth environment offered by the floating zone. As opposed to techniques where the melt is in direct contact with a crucible, such as the Czochralski or Bridgman techniques, the melt only touches materials composed of the desired reagents. This eliminates the diffusion of unknown impurities from crucibles into the melt, and subsequently contaminating the growing crystal. In addition to eliminating this extrinsic source of defects, intrinsic impurities initially present in the polycrystalline material are largely reduced in the grown crystal by the process of zone-refinement, which occurs throughout the growth. As the molten zone is passed along the feed rod, impurities may either tend to be ejected from the growing crystal and accumulate in the melt or vice versa. This behavior is dependent on the segregation coefficient, k, the ratio of the equilibrium solubility of the impurities in the solid to that of the liquid melt. For instance, if $k < 1$, as the crystal solidifies and thermodynamic equilibrium is achieved, it continuously ejects impurities back into the melt [8]. The result is that, in the case that $k < 1$, impurities can be driven to one end of a single crystal boule under repeated zone refinement.

7.2.2. *Growth parameters*

As is the case for any crystal growth process, phase diagram considerations (discussed in Chapter 3) are critical in guiding FZ-growth. While the zone refining process can dramatically decrease the level of foreign elements found in the crystal, impurities in the form of nearby secondary phases with some or all of the same elemental constituents as the desired phase cannot be avoided in the same way. Rather, the different parameters described below must be tuned to stabilize growth of the desired phase. Broadly, compounds for FZ-growth can be separated into *congruently* and *incongruently* melting compounds. In compounds which undergo congruent melting, the composition of the liquid phase is maintained in the crystallization of a single solid phase. As a result, a congruently melting compound can be grown from a feed rod whose stoichiometry is that of the desired compound. For incongruently melting compositions, this is no longer the case and traversing the liquidus typically leads to a two phase region, such as for a compound with a peritectic point [6, 8]. In this regime, variations of the standard FZ-growth technique must be utilized, as are discussed in Section 7.3.

Stability in an FZ-growth is largely determined by the ability to reach a steady-state mass flow through the molten zone. In the steady state, the volume of the zone remains roughly constant, meaning that the incoming and outgoing mass flows are balanced, and the liquid bridge between the feed and growing crystal is adequately stabilized by surface tension. The shape of a stable zone in the steady state is generally symmetric in the axial and radial directions with a roughly square aspect ratio [6]. The mass flow into the melt is controlled by the speed of the feed rod and/or the hot zone, while mass flow out is controlled via the rate of crystal growth. Growth rates found in the literature vary widely depending on the material being grown [4, 6, 8], with rates of .05 mm/hr for $Bi_2Sr_2Ca_2Cu_3O_{10}$ [9] and 240 mm/hr for GaAs [10], at the extremes. Typical rates fall between 1 and 50 mm/hr, with rates in the 1–10 mm/hr range being most common. As a rule of thumb, the slowest stable growth speed improves crystallinity and mosaic (the spread of grain orientations defining a "single" crystal) by allowing for a uniform, close to planar, growth front which is favorable to the grain selection process. In certain cases, a higher growth rate may be needed to deal with thermally induced cracking or volatility (by minimizing the amount of time spent molten) [4].

Most often, the feed and seed rods are counter-rotated during growth, with rates ranging from 5 to 50 rpm [8], in order to maintain a uniform composition of the melt via mixing. Additionally, the feed and seed rotation rates are typically set with a slight differential to prevent destabilizing resonances. Despite this mixing, the molten layer at the liquid-crystal interface can become depleted in one or more of the constituent elements. Based on the phase diagram, this depletion can lead to constitutional supercooling in regions of the melt near the interface, which can cause the growth front to break up and destabilize the molten zone. Both growth rate and rotation rate have been reported to influence the formation of cracks, grain boundaries, and voids as well as the formation of second phases and inclusions, though growth rate is typically the more sensitive variable [4, 6, 8]. Zone stability may also be influenced by composition-dependent properties of the melt itself, since these properties are linked to the surface tension that maintains the zone. For example, high density and low viscosity of the melt may reduce its ability to form a stable molten zone [6]. Sharpening the heating gradient at the boundaries of the hot zone (which can be done by masking the growth chamber or using focused optics, see Section 7.4.3) can help alleviate stability issues [6, 8]. For particularly fragile molten zones increased stability can sometimes be attained by not rotating the samples during the growth process, with the trade-off that mixing of the melt will now rely entirely on the relatively slow process of thermally driven convection.

Control of the composition and pressure of the growth atmosphere is also an important consideration. Depending on the specific chemistry of the material being grown, an oxidizing (O_2), reducing ($Ar + H_2$) or inert (Ar, N_2, He) gas may be utilized to stabilize the growth process. Post-growth analysis often provides clues as to how the growth atmosphere should be modified: for example, formation of an oxygen-rich secondary phase in a crystal grown in air may imply the need for a more oxygen poor growth atmosphere. Most often, a dynamic pressure of the working gas is utilized, with flow rates typically ranging from 0.1 to 1.0 standard L/min, to keep the atmosphere at the melt consistent across the course of the growth. Volatility can also be directly mitigated, to some degree, by the application of an elevated gas pressure, thereby making growth conditions more manageable. Reducing volatility can also improve stoichiometry and phase purity; the effects of pressure are, however, dependent on the specific material system under investigation. High-pressure FZ approaches are discussed further in Section 7.5.

Careful preparation of the feed rod also contributes greatly to the ability to establish a stable molten zone during growth. The precursor material for FZ growth is powder of the desired composition (plus potential flux as discussed in Section 7.3), most often prepared by conventional solid-state synthesis methods. The reacted powder is typically finely ground using ball milling and/or grinding in a mortar and kept dry in order to promote densification during the pressing process. Rods of the material can then be produced by cold pressing within a mold in an isostatic press, followed by a suitable sintering treatment in a furnace. While hot-pressing can produce high density rods, the use of metal or carbon dies, even at low temperatures, can introduce impurities detrimental to delicate quantum states [11, 12]. Sufficient density of the sintered rods, ρ_{rod}, relative to the theoretical density of the crystal lattice (the "X-ray density"), ρ_{xtl}, is important, with ideally $\rho_{rod} > 95\%$ of ρ_{xtl}. Low rod density can lead to the formation of gas bubbles within the molten zone, which can severely destabilize the molten zone, or to dissolving of the feed rod by the melt via capillary action [6]. The value of ρ_{rod} needed for stable growth is, however, a material-dependent value, as is the obtainable ρ_{rod}. Particularly, refractory materials can be difficult to sinter to high densities in typical furnaces, and in certain cases, an initial rapid translation of the feed rod through the hot zone of the FZ furnace can be used to increase density. A further consideration is that the feed rod must be made sufficiently long to allow time for grain selection to occur — typically $\sim$6 mm diameter rods are formed 150–200 mm in length. Care should be taken to make the rods as straight as possible so that they can be aligned coaxially in the furnace, thus avoiding precession of the melt out of the hot zone. A uniform diameter is also desirable to maintain a consistent input mass flow.

7.2.3. *Assessment of the crystal boule*

As mentioned in Chapter 1, the quality of any grown crystal should be carefully assessed by lab-based techniques before it is utilized for more time consuming or costly experiments. While what constitutes crystal quality varies, some general features of a high-quality crystal include: a single lattice orientation (i.e., sharp mosaic), narrow spread in lattice parameters (i.e., good crystallinity), the correct stoichiometry, and the absence of impurity phases or inclusions. It is important to emphasize that the fact of a continuous, stable growth does not, by itself, guarantee a high-quality single crystal. Furthermore, one should keep in mind that sample quality can vary along the length of the boule and in the radial direction. If a polycrystalline seed has been utilized, then the material at the beginning of the growth will certainly be multigrain and the mosaic will likely improve farther along the boule. The phase or chemical composition may also vary along the length, especially in the case of incongruently melting compounds, where the desired phase will likely only be stabilized after significant growth has occurred. The very tip of the crystal, which was cooled or quenched at the end of the growth, is also typically considered not representative of the grown crystal and removed.

Following a growth, powder X-ray diffraction (PXRD) on a pulverized piece of the boule followed by pattern fitting (e.g., Le Bail, Rietveld) is a good first step. This allows for determination of phase purity and lattice parameters, which provides a consistency check for the structure against the precursor powder and literature. Confirmation of the crystallographic structure should be complemented with techniques to determine the stoichiometry of the crystal. Energy dispersive X-ray analysis (EDX), wavelength dispersive X-ray fluorescence (XRF) and electron probe microanalysis (EPMA) can all provide this determination on varying length scales and with differing degrees of accuracy. Other techniques such as thermogravimetric analysis (TGA), or inductively coupled plasma mass spectrometry (ICP-MS) can also provide compositional information.

The microstructure of the grown sample is an equally important question, and it is vital to determine if the sample is actually a *single* crystal or if it possesses multiple, misoriented grains. As a first pass, polarized light microscopy on a polished surface offers a simple way to observe grain boundaries and potential inclusions. Single-crystal XRD (SCXRD) is especially helpful and measurements using a Laue camera can quickly reveal the presence of multiple grains or twin boundaries and can be performed along the length of the boule. SCXRD measurements using a multi-axis goniometer can also be helpful for this purpose and allow one to examine the mosaic of the sample, by collecting rocking curves. In both cases, the relatively short penetration depth of X-rays (10–100 microns depending on sample composition) should be considered in making conclusions from measurements at a single point on the crystal. Other characterization techniques may be employed such as single crystal neutron diffraction to determine the bulk average of the boule's crystallinity and to probe the interior regions inaccessible to X-rays.

7.3. Traveling Solvent Floating Zone and Flux Methods

Many materials decompose prior to melting or form melt compositions with components differing from the desired crystal stoichiometry. Additionally, variable rates of volatilization of constituent elements can inadvertently tune a melt composition away from the desired stoichiometry during the growth process. Materials that exhibit these effects traditionally require the use of a flux or "solvent" to lower the melting temperature of the desired system, to tune to an advantageous position in the phase diagram, or to compensate for the preferential loss of a constituent within the melt during crystal growth. In the following section, we highlight the use of these approaches within the floating zone growth process.

7.3.1. *Traveling solvent, floating zone technique*

For materials that melt incongruently, a flux can often be found which is capable of tuning the stoichiometry of the melt to saturate with the desired phase or to lower the melting point to avoid decomposition. A typical example is that the flux composition is tuned such that a peritectic point can be accessed and the zone temperature held close to this point. As the zone/material translates and the melt cools, the desired solid phase nucleates/crystallizes within the melt and crystals may be grown at the seed growth front. This process, similar in spirit (if not in execution) to the flux/solvent-based methods discussed in Chapters 1, 3, 5 and 7, is implemented in FZ-based crystal growth via the traveling solvent floating zone (TSFZ) method.

The principle of the TSFZ method relies on the fact that the solvent material used has a lower melting point and lower density than the desired phase at the growth temperature. As the crystal is grown (and the target stoichiometry is solidified out of the melt), liquid flux is returned back into the molten zone and polycrystalline feed material is fed in to replenish the target melt composition. Diffusion of the desired solid phase through the melt is typically slower than congruent melting processes and often requires slower growth speeds (e.g., ~1 mm/hr). A good example of this process can be found in the growth of lanthanum copper oxide with a CuO flux [13–15]. The use of a "self-flux" or solvent comprised of the constituent elements of the desired material is highly desired in TSFZ zone and often pseudobinary phase diagrams are used to select judicious compositions for crystal growth (Figure 7.2). Due

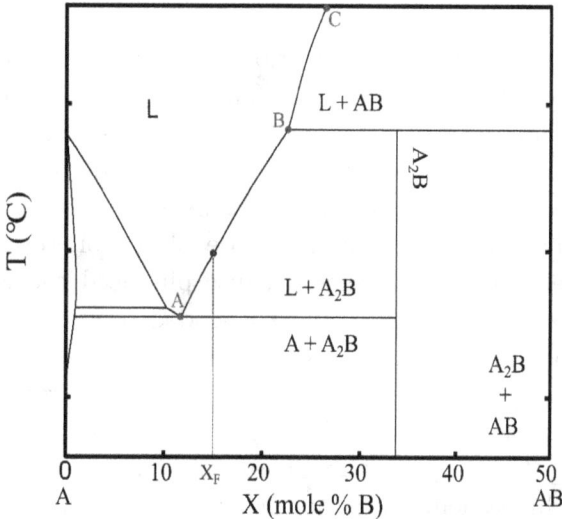

Figure 7.2. Generic pseudobinary phase diagram of compounds A and B illustrating self-flux and self-adjusted TSFZ growth of the line compound A_2B. For self-adjusted growth, a seed rod of A_2B would be melted at $T > T_M$ and the heating power lowered as a rod of A_2B is fed into the zone. By careful temperature adjustment, the liquid composition could be made to move along the line CB to a position on the line BA where steady-state crystallization of A_2B can be established. Alternatively, a disc of A and B, with composition X_F, can be used as a self-flux for growth of A_2B. The self-flux is first melted in the L + A_2B two phase region so that the liquid phase is in equilibrium with A_2B. Steady-state crystallization of A_2B can then be established more quickly by feeding in polycrystalline A_2B.

to its chemical compatibility, the use of a self-flux often mitigates the propensity for extrinsic inclusions of flux material within the crystal boule, where small inclusions of flux material can often strongly renormalize electronic properties in quantum materials [12, 16]. Other factors such as the viscosity of the flux often must also be considered as well as its capability to maintain a stable molten zone throughout the growth process [17]. Chemically inert additives such as B_2O_3 are, for instance, sometimes added to the starting flux composition in order to increase the melt viscosity [18, 19].

The general procedure for TSFZ is to start by cutting a disc of the desired flux material with the appropriate volume (molten volume approximately equal to that of the hot zone) and then placing the disc on top of the crystal seed prior to growth. The hot zone can then be used to first premelt the flux on top of the seed crystal, followed by joining the top polycrystalline feed rod into the melt in order to complete the molten zone. When viscosity is an issue or if the polycrystalline rod needs to be prewet prior to forming the zone, the flux can alternatively be placed on top of the feed rod and premelted. The premelted disc plus feed rod can then be removed from the furnace, hung suspended over the seed crystal, and then joined as normal. A common challenge that often arises in maintaining the molten zone with a flux is the propensity of the flux to "creep" up the feed rod and solvate the rod prior to reaching the melt [13]. Creating a highly dense rod (i.e., >98%) with minimal surface roughness helps mitigate this problem, as well as sharpening the axial heating gradient.

7.3.2. *Flux-feed and self-adjusted growth techniques*

Many materials contain volatile constituents that preferentially evaporate or sublimate during the growth process. As a result, the chemical composition of the molten zone changes as a function of time and the window for growing the desired phase becomes strongly coupled to the growth parameters (seed pull rate, feed rate, starting composition, etc.). Managing the growth of these systems often requires incorporation of a compensating flux into the feed rod itself.

A prominent example of this is the growth of precious metal oxides such as ruthenates [20–22]. An excess of RuO_2 allows the stabilization of the growth zone via compensating for the loss of RuO_2 (volatilized as RuO_3 gas) during the growth process. This approach also requires relatively quick growth rates that help mitigate the rapid loss of material during the growth process; however, not all materials can be stably grown at such rapid pull rates. During the slow growth rates needed for standard TSFZ, this "flux feed" approach may also be used to compensate for the loss of relatively less volatile components of the melt during extended growth runs. An example here is the growth of a number of copper oxides such as La_2CuO_4 where a small excess of CuO is added to the feed rod to replenish the loss of CuO within the CuO-rich flux during the extended growth process [15].

An additional mode of growing incongruently melting materials via FZ-growth is via using "self-adjusted" growth [6, 23]. Rather than adding a starting solvent or flux, for select materials, this method of growing crystals relies on the decomposition of the feed material upon heating to traverse the phase diagram of a compound (Figure 7.2). The idea of this technique is to achieve a window of time where, by adjusting the temperature and growth speed, the desired phase crystallizes out of the melt. Detailed knowledge of a material's phase diagram is required to sustain growth using this technique, although it often has exploratory value in the initial stages of understanding the melt behavior and growth dynamics of a material.

7.4. Types of Floating Zone Furnaces

Development of floating zone-growth instrumentation has been ongoing since the techniques conception in the early 1950s and a variety of floating zone furnaces have been developed in efforts to improve crystal quality, enable growth of different types of materials, and to expand the available processing parameter space. In this section, we discuss typical varieties of floating zone furnaces with an aim towards highlighting the different possible heating sources and geometries.

7.4.1. *Induction melting based floating zone furnaces*

One of the earliest conceptions of a floating zone furnace [1] utilized induction — via Joule heating of a coaxial graphite ring using a radio frequency (r.f.) coil — to produce the narrow hot zone needed for floating zone growth. In the modern conception of an induction-based floating zone furnace, r.f. induction coils produce eddy currents directly in the sample to be melted [24, 25]. The sample is typically enclosed within a quartz growth chamber, with the induction coil on the outside, providing compatibility with pressurized gas atmospheres or vacuum. Eddy current concentrators

[26] can be used to increase the efficacy of the inductive heating of the sample. Different coil designs have been created to allow finer control of the zone size and interface shape [27, 28], as well as to take advantage of electromagnetic stirring of the melt and increase growth volumes [24]. More recently, ultrahigh vacuum induction furnaces, employing metal growth chambers, have also been developed [29]. Further discussion of the induction melting technique, beyond its use in floating zone growth, can be found in Chapter 5.

A notable advantage of induction heating is the provision of relatively straightforward access to high growth temperatures, in excess of 2500°C [26] with more recent growths at up to 3000°C [30], which may be difficult to accomplish for reflective materials in optical furnaces (Section 7.4.2). As such, induction furnaces have extended the advantages of floating zone growth to a variety of high melting point compounds such as rare-earth and transition metal borides [31, 32] and carbides [30]. They have been used to explore binary and ternary rare earth-transition metal silicides [33, 34] and borocarbides [35, 36]. These materials have gained particular interest to the condensed matter community due to observations of unconventional superconductivity [35, 37], Kondo physics and heavy fermion behavior [38, 39]. However, the nature of induction heating means that its use is limited to metallic samples in which significant eddy currents can be created by the electromagnetic field, severely limiting the applicability to oxides and large bandgap materials.

7.4.2. *Lamp and mirror-based floating zone furnaces*

Rather than induction heating, the use of focused optical (infrared) heating makes mirror furnaces extremely versatile crystal growth platforms as both conducting and insulating materials can be grown. The bulk of growth efforts with mirror furnaces have focused on oxide materials, for which optical heating is well suited and the sources of research interest are myriad. While growth of intermetallic compounds with mirror furnaces may encounter problems due to the high reflectivity of densified feed material preventing melting, recent efforts have shown that this obstacle can often be overcome (see Section 7.5) [40].

Conventional optical floating zone furnaces utilize high wattage light bulbs to produce the required heating power. Each bulb is positioned within an ellipsoidal mirror resulting in a high power density transmitted through the mirror's focal point. Several possible arrangements of focusing mirrors exists, but, for each geometry, the mirrors concentrate the light from aligned bulbs onto a single point along the growth axis, forming the hot zone. The hot zone is typically imaged within a quartz or sapphire vessel that is coaxial with the growth direction and allows for the introduction of an isolated and tunable gas environment. Dynamic seals, typically O-rings, maintain a pressurized environment and allow the rotating drive shafts to enter the growth vessel from above and below. Shutters or masks placed in the optical paths between the bulbs and the chamber can be used to further tailor the extent of the hot zone.

The input power range of the furnace can be modified by changing the type of light source used. Incandescent halogen bulbs of varying powers ($\sim$100–1500 W) are a versatile source and can be used to reach melting temperatures typically up to $\sim$2100°C, depending on the optical absorption/reflectivity of the material. While higher wattage bulbs make the growth of higher melting point materials possible, they typically also provide larger starting images due to their larger filaments. This produces a larger hot zone with decreased heating gradients along the growth direction and, in some cases, can decrease the stability of the molten zone. As an alternative, compact xenon arc lamps provide for dramatically higher power densities and allow for growths up to $\sim$3000°C, sufficient to melt highly refractory materials. Unfortunately, the power output of xenon lamps is difficult to control precisely at low powers, often making them unsuitable for low melting point growths. New furnace designs have ameliorated this problem with innovative shutter designs [41].

The most commonly used mirror geometry is the horizontal one, where multiple mirrors and light sources are placed in a circular array perpendicular to the growth direction (see Figures 7.3(a) and 7.3(b)). Commercially available horizontal mirror furnaces utilize either two-mirror (e.g., Canon, Quantum Design, Cyberstar) or four-mirror (Crystal Systems Corporation) designs. Compared to four-mirror designs, the two-mirror design is preferred by some for

(a)   (b)   (c)

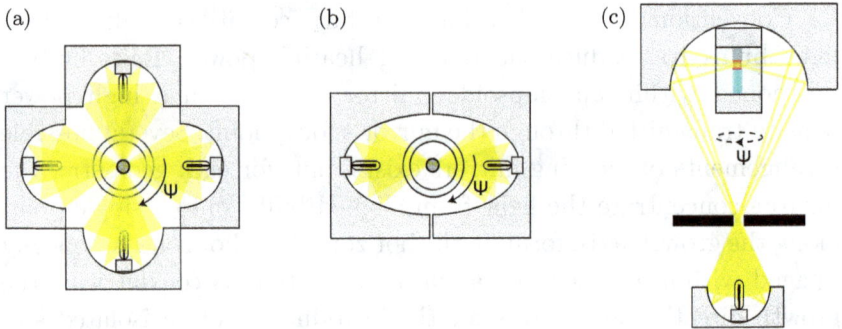

Figure 7.3.   Schematic depiction of the different lamp and mirror geometries discussed in the text. Growth direction is out of the page for the horizontal geometries and within the page for the vertical one. (a) Four-mirror horizontal setup with four lamp/mirror units arranged around the axis of rotation of the growth chamber (ψ). (b) Two-mirror horizontal setup with two lamp/mirror units arranged around the axis of rotation of the growth chamber. (c) Single-mirror vertical setup utilizing a shutter (black rectangles) in addition to a mirror with focal point within the growth chamber.

low melting temperature applications since an accurate alignment of the bulb filaments is easier to achieve. This maximizes the axial heating gradient available from low-power/small-filament halogen bulbs. However, four-mirror designs provide increased heating power and circumferential uniformity, often allowing for larger diameter crystals ($\approx >8$ mm) of higher melting point materials to be more easily grown.

Alternatively, vertical mirror geometries (see Figure 7.3(c)) are also being increasingly utilized [41, 42]. In this geometry, a single lamp and ellipsoidal mirror are placed coaxial with the growth direction, below the hot zone. The light from this lower mirror is focused onto the hot zone using a second ellipsoidal mirror above the lamp. This second mirror provides power input in the same plane as the previously discussed horizontal geometry and can produce extreme circumferential uniformity at high power. This arrangement has seen renewed attention in the field of high-pressure floating zone, as discussed in Section 7.5. Another variant of floating zone furnace design is the horizontal tilting mirror furnace [43, 44]. In this design, the mirrors can be tilted out of the horizontal plane perpendicular to the growth direction. Such a design reduces the effective power due to the reduction in reflecting area of the mirror needed to allow the tilting but allows for tilt values of up to 30 degrees out of the horizontal plane. An increased tilt angle has been shown to alter the shape of the solid–liquid interface by decreasing the convexity of the molten zone [43]. For instance, in growths of rutile-TiO_2 this change in solid–liquid interface resulted in lower defect density and enabled increased sample volume via larger diameter stable molten zones [44].

7.4.3. *Laser-based floating zone furnaces*

The benefits of laser-based floating zone furnaces, where lasers replace lamps and mirrors in geometries similar to those described above, have been recognized and employed in custom setups for nearly 50 years [45, 46]. However, it is only recently that the emergence of relatively affordable kW-level, compact diode lasers has made commercial laser furnace systems available to the general research crystal growth community [47].

The move from incoherent light sources (i.e., lamps) to continuous wave lasers is motivated by several factors. Provided the wavelength of the laser is judiciously chosen, power directly transmitted to the material to be grown can be dramatically increased compared to lamp-based heating [45]. However, one challenge of the inherently narrow line-width of lasers is an increased sensitivity to the optical absorbance and reflectivity profiles of the material to be grown [7]. In most scenarios this can be mitigated, although highly transparent seed crystals and high-density metal feed/seed materials can be challenging to effectively couple to and melt. A further advantage of lasers is the ability to use conventional optics — arranged horizontally around the growth chamber — to produce tunable, highly focused, and homogeneous optical zones, as opposed to shuttering techniques for lamps. This allows for a concentrated and well-defined hot zone to be formed [30]. The compact optical zone formed by the laser light in turn provides a much sharper axial heating gradient than lamp-based furnaces; gradients higher than 1500 °C/cm may form near the solid–liquid boundary of the molten zone [47, 48] compared to $200 - 400\,°C/cm$ found in typical lamp-based furnaces due to the presence of non-normal incidence light from the conventional bulbs [49].

Laser-based heating, in several scenarios, also provides a variety of benefits to the floating zone growth process itself. The most obvious improvement, which motivated the initial development of laser-based floating zone, is the increased power density allowing for increasingly high growth temperatures [7]. This eases the challenge of growing high melting temperature materials, increasing access to interesting refractory materials such as pyrochlore zirconates [50]. The improved focus of lasers offers several advantages in addition to simply attaining higher melt temperatures. Improved circumferential uniformity can be obtained in geometries with a large number of incident laser beams, which helps reduce oscillations in the surface temperature of the melt and avoid subsequent rapid freezing events which can destabilize the zone [47]. The steep axial gradient helps to sharpen molten zone boundaries which can lead to more stable growths than in a mirror furnace, as in the case of $GdTiO_3$ [50]. Importantly, the stability of traveling solvent floating zone growths (Section 7.3) can also be improved in this manner since significant heating outside of the intended hot zone can cause the melt to attack the feed rod and lead to instabilities, as demonstrated in the improved growth of $BiFeO_3$ [47]. The narrower melt zone obtained via lasers also leads to a reduction in the amount of molten material, which is beneficial for the growth of volatile materials [48]. Finally, laser-based heating also provides a reduction in the needed solid-angle of optical access into the growth chamber. As discussed in Section 7.5, this advantage has recently been exploited to increase the growth pressures attainable with the floating zone technique.

7.5. Floating Zone Growth at Environmental Extremes

7.5.1. *High-pressure floating zone*

The motivation for high-pressure floating zone growth stems from the fact that pressure provides a thermodynamic variable for tuning phase stability as well as a method to reduce sample volatility. High-pressure growth has proven an effective way to stabilize sought after phases [51–53] or attain improved properties for compounds previously synthesized at low pressure [50, 54]. The ability of high-pressure FZ-growth to synthesize phases which are metastable under standard conditions has significant implications for condensed matter physics research. By providing an ever-broader range of material platforms, new regimes of exotic electronic and magnetic states can be accessed, and high-pressure FZ-growth presents an opportunity for generating ultrahigh purity crystalline material of long elusive compounds. A perspective on high-pressure synthesis outside of the floating zone technique is given in Chapter 11.

The purpose of pressure as a process variable should be considered for its effects on both the thermodynamics *and* kinetics of the growth process. On the kinetic side, pressure acts to reduce the volatility of a component from the molten zone via a molecular collision-based effect. Due to the wide variation in the temperature dependence of vapor pressure for different elements and molecules, the temperature required to melt a compound may lead to a significant amount of one component being expelled from the melt (i.e., volatilizing). If a sufficiently high pressure of gas exists inside the chamber, then a statistically significant number of collisions between the molecules of the volatile species and the gas molecules will occur at the surface of the melt. Such a reflecting collision has the potential to balance the vapor pressure of a volatile component, and thereby reduce the bulk volatility from the melt that occurs. This intuitive effect is formally described by the kinetic theory of gases from which an expression can be derived that shows a direct dependence of the evaporation rate of a molecule on a constant that is inversely dependent on pressure [55, 56]. Excessive volatility is a common reason that single crystals of a compound are inaccessible by FZ-growth due to problems of suitable power input, zone stability and the ability to image the growth. Reducing volatility by increased gas pressure is also desirable to minimize defect formation and off-stoichiometry in the grown crystal due to depletion of the volatile element in the molten zone. Recently, the pressures obtained in modern high pressure furnaces have begun to surpass the critical pressures of the commonly utilized gases, implying that the growth atmosphere has transitioned to a supercritical fluid (O_2, H_2 and Ar all have critical pressures in the range 10–50 bar and critical temperatures well below room temperature). Currently, minimal quantitative experimental work has been performed to understand the effects of this transition on floating zone growth. However, in certain regimes the growth medium may begin to *promote* volatility in the supercritical regime, as the fluid becomes able to solvate and dissolve the compound being grown [56].

Many phases for which it would be desirable to grow single crystals by FZ are inaccessible at low pressure since their compositions do not lie in congruent melting regions of the phase diagram. This inevitably leads to severe stoichiometry issues or formation of other phases. However, since the free energy landscape of a material system is dependent on all thermodynamic variables, pressure provides a relatively straightforward way to tune the phase diagram and either promote congruent melting or simply stabilize the desired phase. More specifically, pressure allows for direct modification of the chemical potential for ions within the melt thereby allowing the stabilization of different oxidation states which may not be stable at standard conditions. High pressure O_2 environments have proven valuable in this respect, allowing growth of oxide materials which could not be made at low pressures. Notable recent examples include successful single crystal growth of the ferrimagnetic quantum paraelectric $BaFe_{12}O_{19}$ [53] and the correlated metal $LaNiO_3$ [52] at pressures ranging from 40 to 150 bar O_2.

The potential to synthesize materials far out of standard temperature and pressure equilibrium has led to the continued development of furnaces for high-pressure floating zone growth. Early work established the use of xenon bulbs in a single lamp, double-mirror, vertical geometry for high-pressure growths from 80 to 100 bar in thick-walled fused silica or sapphire tubes — made possible by the narrower solid angle required for heating and subsequent shorter growth chambers [41, 42]. In comparison, the most common horizontal mirror furnaces available today typically reach a maximum growth pressure of $\sim$10 bar using fused silica growth chambers.

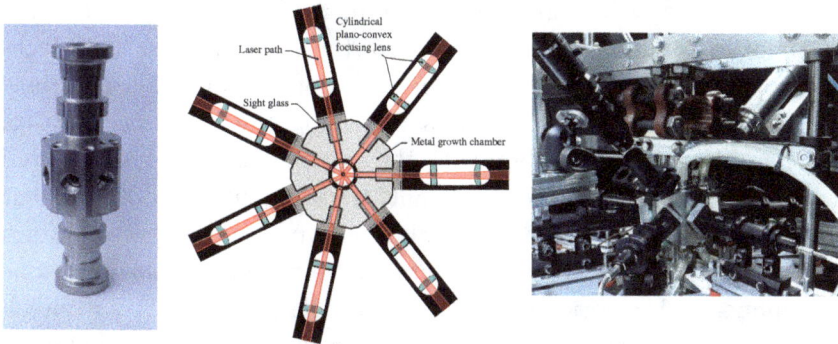

Figure 7.4.   Pictures showing the novel high-pressure floating zone setup at University of California, Santa Barbara [3] — latest all metal growth chamber designed for imaging in plane of the lasers (left), cutaway showing the laser path through the focusing optics and chamber (middle), overview of the current setup, ready for a growth (right).

Currently, Scientific Instruments of Dresden (ScIDre) produces a vertical-mirror furnace capable of providing pressures up to 150 bar in a sapphire growth chamber, and has recently introduced a system capable of performing stable growths in a vertical-mirror geometry at pressures up to 300 bar [56]. More recently, a novel furnace design using laser-based heating has pushed this pressure boundary still further, with successful growths at 700 bar and the capability to attain 1000 bar [50]. In laser-based systems, the reduced optical access required for the tightly focused lasers allows for the use of a high-strength metal growth chamber with discrete windows, enabling much higher growth pressures than transparent chambers (see Figure 7.4). Pushing this pressure frontier further is central to expanding the accessible phase space of the technique, whether by lessening issues associated with volatility or by directly modifying the thermodynamics of the crystal growth. The rapidly expanding field of high-pressure FZ, particularly through the application of laser-based heating, is a promising avenue towards establishing new ultrahigh purity crystal growth regimes.

7.5.2.   *Ultrahigh vacuum floating zone*

Ultrahigh vacuum (UHV) floating zone aims to provide a growth environment that is extremely clean of foreign contamination, allowing for growth in ultrahigh vacuum conditions or the establishment of ultrahigh purity gaseous growth environments. This technique is of great importance in the realm of intermetallic materials, where successful growth requires managing the high reactivity of oxygen, nitrogen and carbon with the starting elements. At elevated temperatures near the melting point, trace amounts of contaminant gasses present in the growth chamber will inevitably react and form second phases or serve as dopants. In terms of the floating zone technique, such reactions can destabilize the molten zone and break up the crystallization front, preventing growth [11]. Although a variety of well-established techniques for the growth of intermetallic compounds exist, such as Czochralski (see Chapter 5) or Bridgman growth, the development of ultrahigh vacuum image furnaces extends the advantages of the floating zone technique into a wider array of intermetallic compounds [40, 57].

While commonly available optical floating zone systems offer the ability to perform growths under high vacuum ($<10^{-4}$ Pa), usually by providing a port for attachment of a vacuum pump to the chamber, they are not optimized for UHV conditions ($<10^{-7}$ Pa). As such, growth of intermetallics by optical FZ, is less common than the growth of oxide materials. Attaining UHV conditions in a commercially available mirror-based image furnace can be achieved via modifications introducing knife-edge metal seals, rather than O-rings, introducing bake-out procedures to remove adsorbates from the chamber walls, and providing for chamber seals based on remote magnetic coupling of shafts or, in less stringent cases, magnetic fluid seals.

The seminal demonstration of this approach is shown in [40] where magnetically coupled, all-metal feedthroughs and specially designed quartz-to-metal seals for the quartz growth chamber were employed. The magnetically coupled feedthroughs eliminate the need for any lubrication of the drive shafts, eliminating a major source of contamination and the "flanged" quartz tube seals by compressing a metal gasket against metal bellows. The nearly all metal construction is compatible with a bake-out procedure and allows UHV conditions to be reached, while the bellows provide mechanical flexibility for accommodation of up to 10 bar of gas pressure. Several successes have been demonstrated using this technique, ranging from Half Heusler magnets to skyrmionic materials [58–60].

Acknowledgments

Eli Zoghlin and Stephen Wilson gratefully acknowledge funding support from the W. M. Keck Foundation.

Bibliography

[1] H. C. Theuerer, Method of processing semiconductive materials, U.S. Patent No. 3,060,123 (23 Oct. 1962).
[2] P. H. Keck, and M. J. E. Golay, *Phys. Rev.*, **89**, 1297 (1953).
[3] P. H. Keck, *et al.*, *Rev. Sci. Instrum.*, **25**, 331 (1954).
[4] S. M. Koohpayeh, D. Fort, and J. S. Abell, *Prog. Crystal Growth Characterization Mater.*, **54**, 121 (2008).
[5] S. M. Koohpayeh, *Prog. Crystal Growth Characterization Mater.*, **62**, 22 (2016).
[6] H. A. Dabkowska, and A. B. Dabkowski, in *Springer Handbook of Crystal Growth* (Springer, 2010), pp. 367–391.
[7] J. L. Schmehr, and S. D. Wilson, *Ann. Rev. Mater. Res.*, **47**, 153 (2017).
[8] H. A. Dabkowska, *et al.*, in *Handbook of Crystal Growth: Bulk Crystal Growth: Part A* (Elsevier, 2014), pp. 281–329.
[9] T. Fujii, T. Watanabe, and A. Matsuda, *J. Crystal Growth*, **223**, 175 (2001).
[10] A. Croll, *et al.*, *J. Crystal Growth*, **166**, 239 (1996).
[11] J. Bobowski, *et al.*, *Condensed Matter*, **4**, 6 (2019).
[12] Z. Q. Mao, Y. Maeno, and H. Fukazawa, *Mater. Res. Bull.*, **35**, 1813 (2000).
[13] C.-H. Lee, *et al.*, *Superconductor Sci. Technol.*, **11**, 891 (1998).
[14] F. Zhou, *et al.*, *Superconductor Sci. Technol.*, **16**, L7 (2003).
[15] S. Komiya, *et al.*, *Phys. Rev. B*, **65**, 214535 (2002).
[16] Y. Su, *et al.*, *Phys. Rev. B*, **79**, 064504 (2009).
[17] H. Roth, Single crystal growth and electron spectroscopy of d1-systems. Ph.D. Dissertation, Universitat zu Koln, (2008).
[18] A. Maljuk, *et al.*, *J. Crystal Growth*, **212**, 138 (2000).
[19] Y.-C. Lai, *et al.*, *J. Crystal Growth*, **413**, 100 (2015).
[20] R. S. Perry, and Y. Maeno, *J.Crystal Growth*, **271**, 134 (2004).
[21] S. I. Ikeda, *et al.* *J. Crystal Growth*, **237**, 787 (2002).
[22] N. Kikugawa, *et al.*, *Crystal Growth Design*, **15**, 5573 (2015).

[23] R. Das, et al., Crystal Growth Design, 16, 499 (2015).

[24] Y. Ikeda, RF induction heating apparatus for floating-zone melting, U.S. Patent No. 4,942,279, (17 Jul. 1990).

[25] K. H. Lie, J. S. Walker, and D. N. Riahi, J. Crystal Growth, 100, 450 (1990).

[26] R. W. Johnson, J. Appl. Phys., 34, 352 (1963).

[27] R. Hermann, et al., J. Crystal Growth, 275, e1533 (2005).

[28] A. Mühlbauer, et al., J. Crystal Growth, 151, 66 (1995).

[29] S. Takaki, et al., Le J. Phys. IV, 5, C7 (1995).

[30] S. Otani and T. Tanaka. J. Less Common Metals, 82, 63 (1981).

[31] S. Otani, et al., J. Solid State Chem., 154, 238 (2000).

[32] K. Takahashi, and S. Kunii, J. Solid State Chem., 133, 198 (1997).

[33] G. Graw, et al., J. alloys Compounds, 308, 193 (2000).

[34] G. Behr, et al., J. Crystal Growth, 237, 1976 (2002).

[35] G. Behr, et al., Crystal Res. Technol. J. Exper. Industrial Crystallography, 35, 461 (2000).

[36] G. Behr, et al., J. Crystal Growth, 198, 642 (1999).

[37] R. J. Cava, et al., Nature, 367, 146 (1994).

[38] R. Mallik, and E. V. Sampathkumaran, J. Magn. Magn. Mater., 164, L13 (1996).

[39] R. A. Gordon, et al., J. Alloys Compounds, 248, 24 (1997).

[40] A. Neubauer, et al., Rev. Sci. Instrum., 82, 013902 (2011).

[41] D. Souptel, W. Löser, and G. Behr, J. Crystal Growth, 300, 538 (2007).

[42] A. M. Balbashov, and S. K. Egorov, J. Crystal Growth, 52, 498 (1981).

[43] Md A. R. Sarker, et al., J. Crystal Growth, 312, 2008 (2010).

[44] S. Watauchi, et al., J. Crystal Growth, 360, 105 (2012).

[45] D. B. Gasson, and B. Cockayne, J. Mater. Sci., 5, 100 (1970).

[46] K. Dembinski, et al., J. Mater. Sci. Lett., 6, 1365 (1987).

[47] T. Ito, et al., J. Crystal Growth, 363, 264 (2013).

[48] P. Telang, et al., J. Crystal Growth, 507, 406 (2019).

[49] S. M. Koohpayeh, et al., J. Crystal Growth, 311, 2513 (2009).

[50] J. L. Schmehr, et al., Rev. Sci. Instrum., 90, 043906 (2019).

[51] N. Wizent, et al., J. Crystal Growth, 318, 995 (2011).

[52] J. Zhang, et al., Crystal Growth Design, 17, 2730 (2017).

[53] H. B. Cao, et al., APL Mater., 3, 062512 (2015).

[54] G. Behr, et al., Crystal Res. Technol. J. Experimental Industrial Crystallography, 40, 21 (2005).

[55] I. Langmuir, Phys. Rev., 2, 329 (1913).

[56] W. A. Phelan, et al., J. Solid State Chem., 270, 705 (2019).

[57] A. Bauer, et al., Rev. Sci. Instrum., 87, 113902 (2016).

[58] A. Regnat, et al., Phys. Rev. Mater., 2, 054413 (2018).

[59] T. Schulz, et al., Nature Phys., 8, 301 (2012).

[60] F. Jonietz, et al., Science, 330, 1648 (2010).

Chapter 8

High-Throughput Methods in Superconductivity Research

Jie Yuan[*,†,¶], Valentin Stanev[‡], Chen Gao[§,††],
Ichiro Takeuchi[‡,∥], and Kui Jin[*,†,**]

[*]*Beijing National Laboratory for Condensed Matter Physics,
Institute of Physics, Chinese Academy of Sciences,
Beijing,100190, China*
[†]*Key Laboratory of Vacuum Physics,
School of Physical Sciences, University of
Chinese Academy of Sciences, Beijing, 100049, China*
[‡]*Department of Materials Science and Engineering,
University of Maryland, College Park, MD 20742, USA*
[§]*Beijing Advanced Sciences and Innovation Center of
Chinese Academy of Sciences, Beijing 101407, China*
[¶]*yuanjie@iphy.ac.cn*
[∥]*takeuchi@umd.edu*
[**]*kuijin@iphy.ac.cn*
[††]*gaochen@ucas.edu.cn*

8.1. Introduction

Ever since first observed by H. Kamerlingh-Onnes in 1911 at Leiden, the phenomenon of superconductivity has been intensely studied. The nature of this quantum phase was explained conceptually only by the Bardeen–Cooper–Schrieffer (BCS) theory, developed in 1957. For conventional superconductors, the isotope effect clearly established that the superconducting critical temperature (T_c) is linked to the coupling between the electrons and the lattice, with the strength of the electron–phonon (el–ph) interaction being a key physical parameter. This put a perceived ceiling of 30–40 K for the T_c the standard inorganic metallic crystals [1]. However, in 1986 a ceramic compound Ba–La–Cu–O was found to show superconductivity at 35 K by Bednorz and Müller in Zurich [2]. Soon, the limit of 40 K was crossed [3, 4], and critical temperatures above that of liquid nitrogen (77 K) were observed. At ambient pressure, the highest T_c reached was 138 K in $Hg_{0.8}Tl_{0.2}Ba_2Ca_2Cu_3O_{8.33}$ [5]. These ground-breaking results inspired an intense research effort, yet the mechanism of superconductivity in this class of materials remains unclear.

Before the discovery of high-T_c materials, the majority of known superconductors were composed of no more than two elements and were mostly elemental superconductors, alloys, and intermetallics. As the superconducting critical temperature rose above 40 K, the super-conducting compounds became more complex and tended to contain more elements; for example, $YBa_2Cu_3O_{6+\delta}$, $HgBa_2Ca_2Cu_3O_{8+\delta}$ and $Hg_{0.8}Tl_{0.2}Ba_2Ca_2Cu_3O_{8.33}$ are made up of four, five and six elements, respectively. The discovery of Fe-based [6, 7], Cr-based and Mn-based superconductors [8–10] demonstrated that even elements whose presence was once regarded as strongly detrimental to supercon-ductivity have to be considered. Obviously, the number of possible combinations grows exponentially as more elements are included in potential compounds. The number of compounds can be estimated within the group of natural elements; then, the order of magnitudes for possible combinations are 10^3, 10^5, 10^7 and 10^9 for binary, ternary, quaternary and penternary compounds, respectively. In addition, many methods to modulate superconductivity are known: ultra-high pressure [11, 12], ultra-thin film deposition [13], superlattice architecture [14] and ionic liquid/solid gating [15, 16] being few of those. Third, some high-T_c superconductors are extremely sensitive to physical and chemical parameters. For instance, one percent variation in oxygen content can turn a copper oxide superconductor into an insulator [17]. One will have to do a vast amount of synthesis to construct a reliable and detailed phase diagram as a function of anion content. What is more, a complete phase diagram of high-T_c superconductor must necessarily be multivariate, i.e., must include synthesis conditions, pressure, magnetic field, and other variables. It is clearly not feasible to efficiently construct such a multidimensional phase diagram using the traditional one-material-at-a-time experimental methodology.

Thus, in general there are two challenges to the experimental studies of superconductivity: (i) searching for superconducting candidates in an enormous composition space comprised of more and more elements; (ii) delineating the key physical parameters that control the superconductivity, e.g., by constructing reliable multidimensional phase diagrams. New paradigms and methods are required to boost the efficiency of superconductivity research [18, 19].

In bioinformatics and the pharmaceutical industry, similar challenges arose before the discovery of high-T_c superconductors. There is a huge number of possible gene combinations and drug formulas, and it was not realistic to synthesize and test them one by one. The idea of doing many tests at the same time — the so-called high throughput strategy — was rapidly developed and put into practice. Ever since, the high-throughput methods have been a key driving force boosting up the development of modern biology and medicine. This methodology was also adopted in condensed matter physics and materials science. In 1970, Hanak introduced the rudiment of a multiple-sample-concept in materials research, e.g., by developing a multiple-target, radio-frequency co-sputtering deposition [20]. However, due to the difficulties in obtaining and processing a large amount of experimental data, high throughput methods were not widely adopted in materials research at that time. Another factor inhibiting their spread was the low demand for complex functional materials. This has changed in the last decades, owing to the growing needs of industries such as communications, aerospace, energy, and transportation for novel functional materials. This necessitated significant speeding up of material R&D. High-throughput synthesis and characterization techniques became major research tools, leading to a substantial progress in a number of applications (see, for example, [21–28]). Helping these advances, the rapid progress in information technologies created powerful tools able to deal with the massive amounts of data generated by this approach.

Thus, high-throughput paradigm provides an extremely promising approach to meet the challenges in superconductivity research outlined above. In one early pioneering work demonstrating the potential of these methods, Xiang *et al.* reported the rapid synthesis of high-T_c superconductors, with a 128-member library of copper oxide superconducting thin films deposited on a single substrate in one batch [29]. Although the superconductors synthesized in this work had been known already, the efficiency achieved in the parallel synthesis was encouraging. In 2013, Jin *et al.* provided the first example of a superconductor discovered via the high-throughput methodology. They fabricated composition-spread films comprised of Fe and B, across 3-inch Si wafers by a co-sputtering technique. Indications of a superconducting region were found from the resistance measurement by a homemade probe with an array of 64 pogo-pins [Figure 8.1(b)]. Superconductivity was then confirmed by a zoom-in magnetotransport and susceptibility measurements [30]. This actually became the first example of a superconductor discovered by a high throughput paradigm. In 2016, Wu *et al.* studied

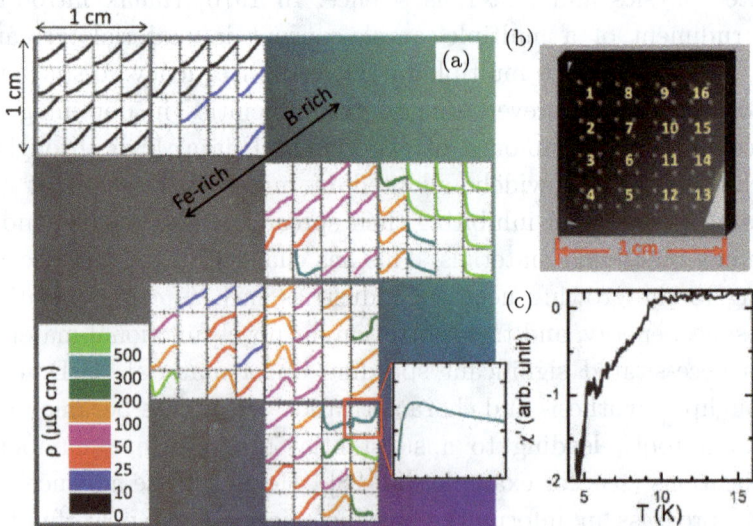

Figure 8.1.   Mapping of the temperature dependence of resistivity on the Fe–B composition spread film.

a library of heterostructures made up of La_2CuO_4 and combinatorial $La_{2-x}Sr_xCuO_4(0.15 \leq x \leq 0.47)$, achieved by a gradient deposition rate of the Sr content in their advanced oxide molecular beam epitaxy system [31]. In 2018, Stanev *et al.* filtered more than 100,000 compounds and created a list of potential superconductors, using a machine learning models trained on the critical temperatures of more than 12,000 known superconductors [32].

As these examples demonstrate, the high-throughput paradigm is already starting to permeate superconductivity research. The general high-throughput materials research procedure has the following steps: sample synthesis, characterization, analysis of the data. Significant increase of the efficiency necessitates the acceleration of every step, as well as substantial coordination among them. In this chapter, we outline the advances in all parts of the paradigm of high-throughput superconductivity research: high-throughput synthesis, high-throughput characterizations, high-throughput theoretical calculations, and machine learning. We also discuss the need to develop the next generation of facilities, techniques, and platforms in order to fully utilize the power of high-throughput methods.

8.2. High-Throughput Synthesis of Superconductors

High-throughput synthesis is one of the foundations of the recent acceleration of the rate of materials exploration. However, to achieve maximum efficiency, common components and procedures used in different experimental steps, like evacuation, heating and conditioning atmosphere, should be combined as much as possible. This is the basis of the so-called combinatorial approach, which is the most efficient high-throughput synthesis method. Koinuma *et al.* gave the basic concept of combinatorial chemistry for solid-state materials in [33]. During synthesis, a chemical reaction for the target material is influenced by many factors. The search for new materials can be regarded as a process of scanning certain points in the phase diagram as a function of multiple variables. Compared to classical synthesis process of effectively scanning point by point, combinatorial approach can quickly map entire regions of key parameters such as composition, temperature, and pressure, responsible for optimizing a functionality.

Thin-film preparation technology is the most widely used combinatorial synthesis method [24]. As early as 1965, Kennedy et al. reported a method for rapidly obtaining a Fe–Cr–Ni ternary library by electron beam co-evaporation of three precursors (Fe, Cr and Ni targets), on an equilateral-triangle metal foil with a side length of 10 inches [34]. Since then, there have been three generations of combinatorial film growth techniques. The first generation realizes a gradient chemical composition by ablating the precursors simultaneously and mixing them using natural diffusion. Co-evaporation [34, 35], co-sputtering [20, 36, 37] and co-laser ablating [38] techniques are all categorized into this group. Figure 8.2 shows a typical configuration of a co-sputtering system. Different sources are mounted against the wafer at a certain angle to the normal direction of the wafer surface. During the deposition, such configuration results in a spatial variation of the deposition rate and allows plumes from different targets to overlap and form a natural composition gradient. Recently, Jin et al. obtained an Fe–B binary spread by such method, with a continuous composition across a 3-inch Si wafer with a 200 nm SiO_2 layer on top [30]. The lower part of Figure 8.2 shows the photograph of one composition spread wafer taken under natural light; the average composition at different positions (solid circles)

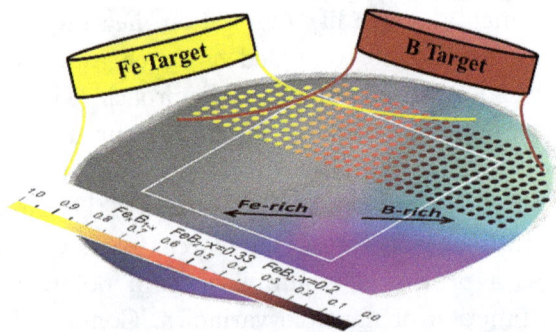

Figure 8.2. Sketch of the co-sputtering deposition for an Fe–B composition spread combi-film.

is obtained by the wavelength dispersive spectroscopy (WDS). The advantage of co-deposition method is that it requires relatively simple facilities and the deposition process is easy to control. However, the chemical composition cannot be precisely manipulated because of the natural distribution of spatial deposition rate. The local composition of the film can only be determined by micro-region analysis methods (discussed in the section of high-throughput characterizations).

To overcome the shortcomings of the co-deposition, Xiang *et al.* developed a method combining thin film deposition and physical mask techniques for a parallel synthesis of spatially addressable libraries of materials [29]. The programmable mask is the core of the second-generation technology. The fabricated films contain discrete squares with various compositions. Because these small regions are addressable, such films are also called integrated materials chips or combinatorial materials libraries. More details about such synthesis method can be found in earlier literature (e.g., [26, 39–41]). In fact, the mask is not necessarily a physical one. Wang *et al.* developed a combinatorial robot which is able to create thick-film libraries by ink-jet printing [42]. After being ground into sufficiently fine powder, precursor can be made into ink [43], taking the role of the three "primary colors." Different doses of inks are ejected on designated places of the wafer. Thus, using selected precursors and an appropriate post processing, an addressable materials library can be crated. In this way, the ejector together with the robot arm replaces the physical mask [44, 45].

Mask pattern has been widely used in study of electronic, magnetic, optical and dielectric materials, as well as catalysts and alloys [46]. However, for a precise study of material properties, distinct techniques such as atomically controlled layer-by-layer thin-film growth are required [19].

Combinatorial laser molecular beam epitaxy (CLMBE) — the third generation of combinatorial thin-film preparation technology — was developed to carry out parallel fabrication via an atomic layer-by-layer process [19]. The state-of-the-art laser molecular beam epitaxy in combination with mobile mask technique [48] has already been used to control layering sequences of high-quality

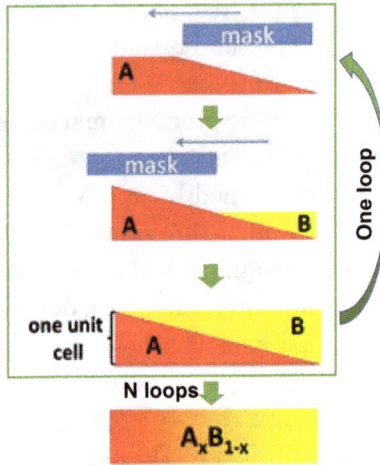

Figure 8.3.   Schematic of the binary-combinatorial-film growth using the continuous moving mask technique [47].

superconducting films. The procedure for preparing binary combinatorial films with continuous chemical composition spread on single substrate can be briefly described as follows (see also Figure 8.3): once a target A is ablated by laser pulses, a metal mask is moved along the substrate in half a period time and results in a linear distribution of the A component. In the other half period, another target B is ablated for a reverse distribution by moving the mask in the opposite direction [47]. A desired thickness of the combinatorial film can be achieved by setting corresponding periods of deposition. It should be emphasized that the two precursors A and B must be mixed in one unit cell during the period, monitored by the reflection high-energy electron diffraction system for *in-situ* diagnostics. Otherwise, a superlattice rather than a combinatorial film will be obtained. For a precise phase diagrams of materials like oxide superconductors [2], CLMBE is easy to operate, and benefits from a wide range of chemical compositions in one batch of deposition. For instance, Yu *et al.* fabricated combinatorial $La_{2-x}Ce_xCuO_{4\pm\delta}$ thin film with x from 0.1 to 0.19 on 1 cm $SrTiO_3$ substrate by the CLMBE [47].

Following the same principle, a ternary-combinatorial film can be grown with alternatingly ablating three targets and shadowing

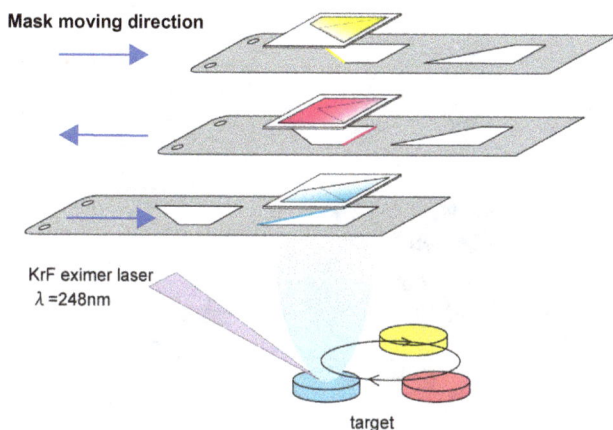

Figure 8.4. Schematic of ternary-combinatorial film growth using the continuous moving mask technique. The blue arrows indicate the direction of mask motion in every step. (From the introduction brochure of Pascal Co., Ltd in Japan).

the substrate from three directions by a mask with sophisticated design pattern as shown in Figure 8.4. With this method, Mao *et al.* prepared an Mg–Ni–Al ternary thin-film library [18]. In combinatorial films, the chemical composition is locked to the fixed space point of the sample surface. For a ternary compound, the formula can be written as $A_x B_y C_{1-x-y}$, in which A, B and C stand for different elements. Since there are two independent coordinates (e.g., x and y) to pin down the space point of one corresponding chemical composition, the full ternary library can be obtained based on single combinatorial spread. However, if more than two variables are involved, e.g., three variables in $A_x B_y C_z D_{1-x-y-z}$, one cannot obtain all the combinations of these four components from a single chip, instead, a pseudo-multicomponent materials library has to be created.

As seen above, three generations of combinatorial-film preparation techniques have been developed; it has to be noted that each generation has its own distinct merits, and is suitable for different kinds of materials. Sometime, a customized combinatorial approach is required, for example for compounds that only show superconductivity in a very narrow range of chemical doping. A typical iron-based

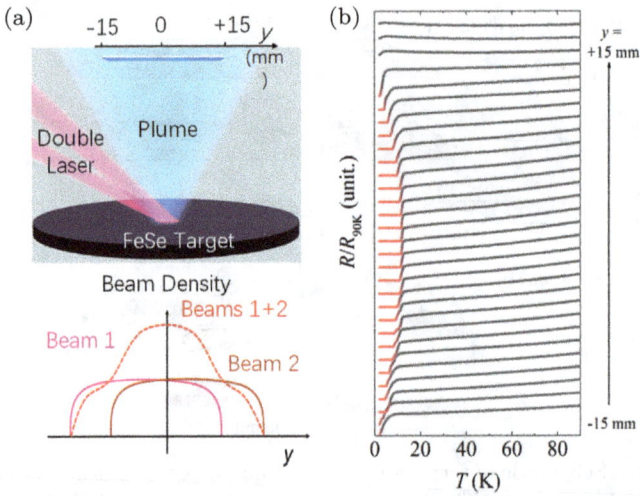

Figure 8.5.   (a) Schematic of the process of double-beam pulsed laser deposition, focused onto the target with controllable displacement. (b) The temperature dependence of normalized resistance along the y direction, with the highest T_c in the middle [51].

superconductor, $FeSe_{1-\delta}$ only displays superconductivity within a finite δ [49]. Since Se is a volatile element, fine control of the Fe/Se ratio of the final film can be extremely challenging by the methods presented above. Recently, a double-beam laser skill was used to create a combinatorial laser fluence on a FeSe target. As reported in the study of $SrTiO_3$ film grown by PLD, the stoichiometric ratio can be regulated by the power density of the laser [50]. Tuning the parameter of the combinatorial laser, the Fe/Se ratio of the final film can be varied in a very narrow δ range (i.e., the order of magnitude is ~0.001 beyond the resolution limit of chemical analysis methods). The left panel of Figure 8.5 shows the schematic map of double-beam laser used in the deposition; the corresponding evolution of the superconducting critical temperature can be clearly seen in the normalized R–T curves from different space regions of the final film (right panel). Moving along the film from one edge to the other, T_c changes continuously from below 2 K to 12 K and back to below 2 K. There is no obvious difference in the film thickness (~150 nm) across the whole film (verified by SEM), whose influence on T_c can be thus excluded. Such combinatorial films offer a good platforms to investigate the nature of tunable superconductivity in the Fe–Se binary system [51].

Wu *et al.* used a tilted Knudsen cell in an oxide molecular beam epitaxy system to synthesize combinatorial $La_{2-x}Sr_xCuO_4$ film. This in principle falls into the class of first generation combinatorial technique, but provides a much finer control over the composition. The deposition rate can be precisely tuned and the growth process can be monitored *in-situ*. The composition gradient originates in the angle between the sources and the substrate. Using this method the composition across the film can be confined to extremely fine steps, $\Delta x \sim 0.00008$ [52, 53].

In addition to generating composition-spread materials libraries, combinatorial approaches can also be used to rapidly ascertain the optimal growth condition for superconducting materials. For example, the conventional way to find the best deposition temperature of a superconducting material is to try different temperatures one by one. However, a much more efficient method is to create a thermal gradient on the substrate during the film deposition [54]. Figure 8.6 shows a configuration that can create temperature gradient during the growth. One end of the substrate is attached to a heater while the other end is kept free-standing. When the heater is set at high temperature, the heat diffuses to the free-standing end through the substrate. Using such configuration, a temperature range in which only pure β-FeSe phase forms was found; only three batches of samples were tested with deposition temperature from 350°C to 700°C.

In addition, a parallel synthesis can also reduce the uncertainty in growth condition between different experimental batches, at least to some extent. Figure 8.7 shows the relationship between the out-of-plane crystal lattice parameter (c-axis) and the T_c extracted from more than one thousand pieces of uniform FeSe films (upper panel) as well as one combinatorial film (lower panel), respectively. It took almost three years to prepare and characterize the 1000+ uniform films in contrast to several weeks for the combinatorial film. Even

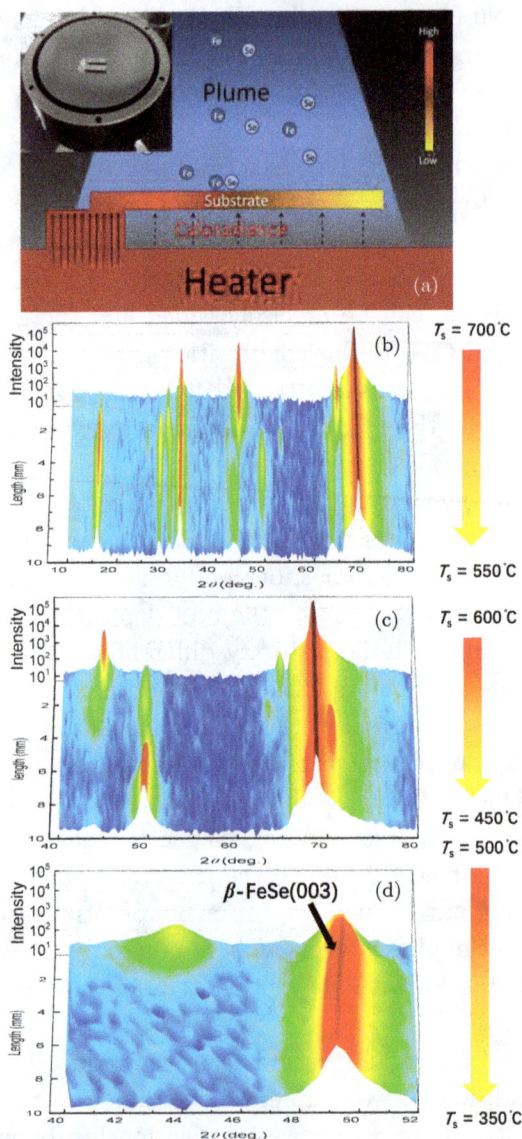

Figure 8.6. (a) Schematic of the combinatorial film growth with gradient temperature, inset on top left is a photograph of the substrate mounted on the heater. (b)–(d) The micro-region $\theta/2\theta$ X-ray patterns of the FeSe films grown with different temperature gradients: 700–550°C, 600–450°C and 500–350°C, respectively. Pure β-FeSe phase can be found in the region where the deposition temperature is not beyond 450°C.

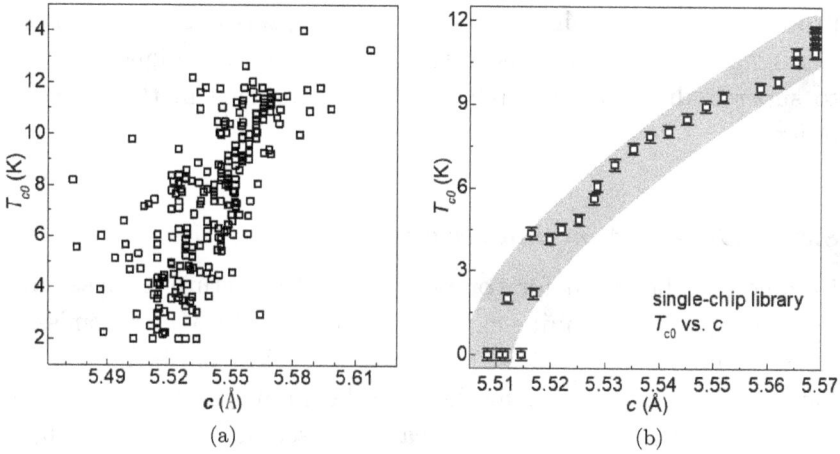

Figure 8.7. The c-axis lattice constants dependence of T_{c0} for (a) uniform FeSe films, (b) combinatorial FeSe film with gradient T_c.

though the amount of data collected from the uniform films is much larger, the positive relation between c and T_{c0} is manifested much more clearly in the combinatorial film data.

Hence, the high-throughput synthesis methods are really effective in extracting the information of superconducting materials, not only on the exploration and key parameters but also the performance of superconductors in different scenarios.

8.3. High-Throughput Characterization Probes

As another step in the high-throughput methodology, rapid characterization of combinatorial films is also indispensable in the drive to accelerate superconductivity research. Owing to the nature of combinatorial synthesis, suitable characterizations probes need to have spatial resolution capabilities. Several commercially available probes satisfy this requirement; atomic force microscope, scanning tunneling microscope (STM), optical microscope, scanning electron microscope are few examples. These commercial tools in auto scan functionality can have high spatial resolution, and are equipped with tips or aggregated beams. There are also diverse homemade probes designed for investigating various properties of the high-throughput samples. In this section, we focus on techniques relevant to superconductivity research, briefly summarized in the following order.

8.3.1. *Composition and structure*

Common probes used for composition and structure analyses are: X-ray diffraction; scanning electron microscope (SEM); transmission electron microscope (TEM). Because the electron beam can be focused, the methods using electron beam naturally provide the required spatial resolution and can be used for high throughput characterizations [27, 55].

Great efforts have been dedicated to developing X-ray-related characterization methods with a spatial resolution. Since X-ray has much higher frequency and photon energy than visible light, they tend to penetrate or get absorbed in most materials. However, most of the techniques used to redirect X-rays are unable to produce well-focused beams.

Initially, pinholes or slits made of anti-radiation materials were used to shrink the profile size of X-ray beams. Figure 8.8 shows the data collected from a combinatorial $La_{2-x}Ce_xCuO_4$ library chip (an electron-doped copper oxide superconductor, x varies from 0.10 to 0.19). Micro-area scan is realized by adding a narrow slot and a moving sample stage to a commercial X-ray diffractometer [47]. The calculated c-axis lattice constant monotonically decreases with raising the nominal doping level x, in accordance with the tendency extracted from uniform LCCO thin films fabricated by the conventional pulsed laser deposition (PLD) technique [56, 57].

In this technique, the analysis sensitivity and accuracy are unavoidably reduced due to the limited incident X-ray throughput, which necessitates time-consuming repetitive angular-scanning measurements. Thus, techniques using a zone plate as an X-ray focusing component or a 2D detector were developed in order to enhance the intensity of the X-ray beam [58] or measurement efficiency [59]. In 2005, Luo *et al.* developed a combinatorial X-ray analysis system

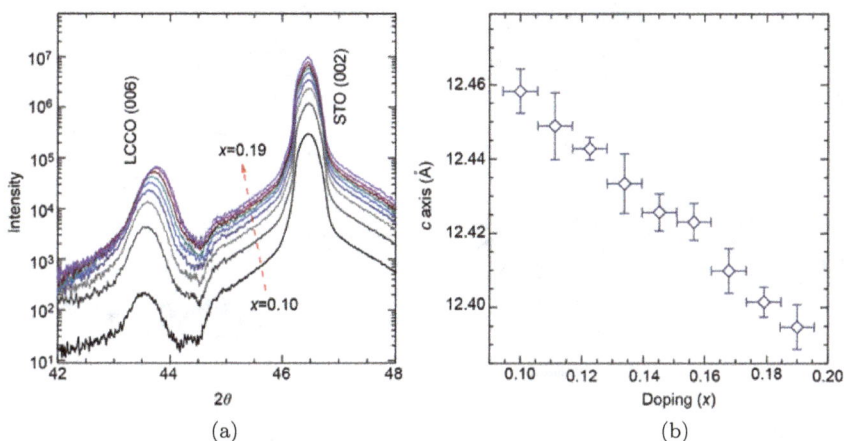

Figure 8.8. (a) The micro-region X-ray diffraction results of combi-film $La_{2-x}Ce_xCuO_{4\pm\delta}$. The component interval Δx is about 0.012. (b) The variation of c-axis lattice constant with increasing doping levels (x refers to nominal doping level).

with a spot size of 0.5 mm by using polycapillary X-ray lens to focus the divergent beam from an X-ray source. This integrated micro-region X-ray diffraction, fluorescence, and photoluminescence system can meet requirements of high-throughput structure, composition and scintillation characterizations. Instead of the angular-dispersive X-ray diffraction, they used energy-dispersive XRD (EDXRD) to utilize the entire spectra of the focused X-ray varying from 4 to 19 keV, free of moving parts. Subsequently, the measurement process was further accelerated. The configuration used in the system is shown in Figure 8.9 [60]. More details about the operation principle of X-ray optics can be obtained in [61].

Concurrently, special X-ray sources were developed to meet the requirements of high-throughput characterizations. Xiang *et al.* used a micro-beam X-ray system to simultaneously screen the composition and the structure with a resolution of 10–100 μm. The X-ray intensity in such system is 20 times higher than that of an ordinary X-ray source [26].

Since synchrotron facilities can offer high spatial resolution while providing adequate photon flux, several high-throughput screening

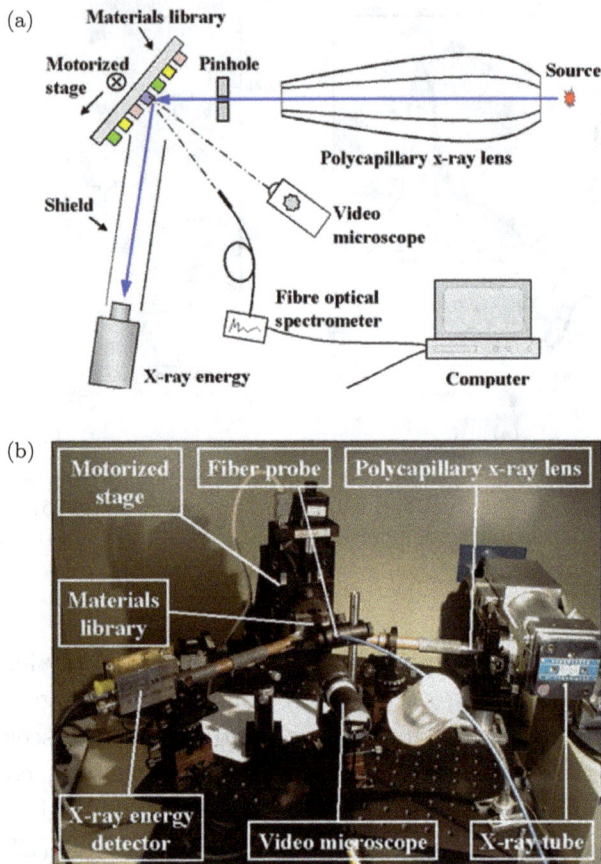

Figure 8.9.   (a) Configuration and (b) photograph of the X-ray high-throughput characterization system.

techniques were developed specifically for such sources. In 1998, Isaacs *et al.* used synchrotron X-ray micro-beam techniques to verify the feasibility of high-throughput characterization of combinatorial libraries [62]. This work included X-ray fluorescence, diffraction and absorption spectroscopy experiments with the spatial resolution of $3 \times 20\,\mu m^2$. In 2014, combinatorial phase mapping techniques with a synchrotron source were realized by Gregoire *et al.* [63]. The technique used in this work provides the capability of measuring more than 5000 samples per day using a synchrotron X-ray diffraction and fluorescence for rapid characterization. Xing *et al.* recently pushed the efficiency and automation even further [64]. The rate of recording diffraction images reached one pattern per second and the diffraction patterns were automatically processed to create the composition-phase map. Apart from the prevalent characterization methods such as diffraction and fluorescence, X-ray absorption is also suitable for high-throughput characterization. For instance, Suram *et al.* used custom combinatorial X-ray absorption near edge spectroscopy techniques to obtain detailed chemical information of samples [65].

Although synchrotron X-ray facilities are powerful characterization tools, they remain a scarce resource, which impedes their use in high-throughput characterization. Developing more accessible rapid screening systems remains an urgent task.

8.3.2. *Transport properties*

Transport properties such as resistivity, magnetoresistance and Hall effect are extremely important in illuminating the nature of superconductivity in various materials.

In 2005, Hewitt *et al.* developed a 196-pin device to measure the *dc* resistance versus temperature as shown in Figure 8.10. The distance between nearest pins is about 4.64 mm. In the configuration of Van der Pauw method, these pins can establish 49 parallel channels in an epoxy plate of 4.2 inches × 4.2 inches. For low temperature measurement, the device can be cooled down to 7 K by cold fingers [66].

Typically, the sample chambers of most commercial low-temperature measurement system, especially the ones equipped with magnet, have limited space. For high-throughput magnetoresistance measurement, a more compact probe device compatible with existing system has been highly desirable. Jin *et al.* designed a high-throughput probe to search for superconductivity in the Fe–B composition spreads mentioned in the previous section. They used pogo pins for a better contact between the sample surface and the pinpoint, and also shorted the distance between nearest pins to ∼1 mm. Thus, even if the lateral size of the sample chamber is

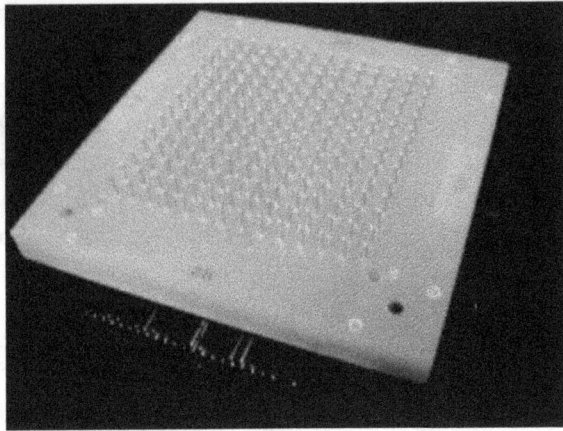

Figure 8.10.   Image of the 196-pin device which makes electrical contact to the 49-sample thin-film library for four-contact resistivity measurements.

limited to a centimeter scale, at least 16 multiplexed channels of temperature-dependent resistance can be measured at 4×4 evenly spaced $1\,mm \times 1\,mm$ areas in the Van der Pauw configuration. Utilizing this probe, Jin *et al.* discovered a superconducting region with T_c reaching 4 K in the region where the ratio of Fe to B reaches 1:2. The temperature-dependent resistivity curves and sketch diagram of the pogo pin array are shown in Figure 8.1. A number of diced $1\,cm \times 1\,cm$ chips from this transition region were measured using the 16-spot simultaneous resistance versus temperature measurement setup. Later, the device was upgraded and made compatible with a commercial physical property measurement system (PPMS). In this way, magnetoresistance and Hall data of a combinatorial film can be rapidly collected.

Bozovic group performed multiple-channel measurements of the Hall signal and the resistivity by a high-throughput system in order to characterize highly conductive oxides with Hall coefficients as small as 10^{-10} m^3/C. The receptacles of this device are shown in Figure 8.11(a). There are 64 slots matching pogo pins with $1.11\,mm$ distance between the adjacent pins, which contact with the patterned film as shown in Figure 8.11(b). The center of the film, a stripe of $300\,\mu m$ in width and $10\,mm$ in length, is bridged with 64 square

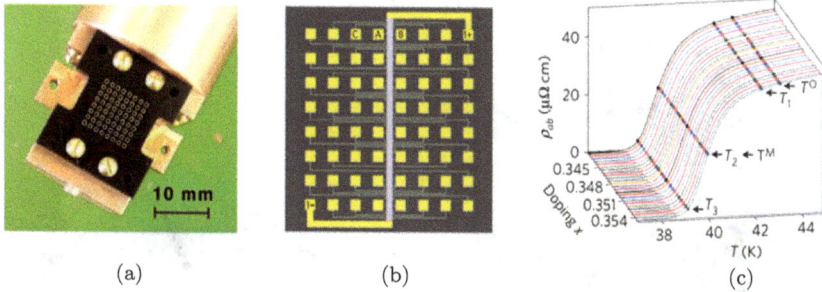

Figure 8.11. (a) Receptacles for spring-loaded pogo pins on the cryostat tail. (b) Film lithography and contact pattern. (c) The temperature-dependent ab plane resistivity ρ measured from different channels on the same $La_{2-x}Sr_xCuO_4 - La_2CuO_4$ bilayer films.

pads covered with gold. The blocks for different channels segment the main stripe to $300\,\mu m$ in length. Using this device, 30 channels for the resistivity and 31 channels for the Hall resistivity can be measured simultaneously. Figure 8.11(c) shows a series of longitudinal resistivity $\rho(T)$ curves from a single $La_{2-x}Sr_xCuO_4$–La_2CuO_4 bilayer films. The doping resolution can reach $\Delta x = 0.0002$ [67]. Due to the high special resolution of the Hall measurements, an intrinsic inhomogeneity of sample could be detected. Wu *et al.* studied the phase diagram and discovered a quantum charge-cluster glass state that competes with the superconducting state [31].

Since the current bridge is shared by all the channels, the sample should have good electrical conductivity. Thus, the pattern is not suitable for screening a combinatorial film with insulating areas. In such case, an array micro-bridge design with independent current paths should be employed. In 2017, Wu *et al.* reported a "sunbeam" lithography pattern for spontaneous Hall measurement [68]. The 36 Hall bars, each with three pairs of transverse contacts (as sketched in Figure 8.12(a), form a circle with 10° between two successive bars. Angular dependence of ρ_T and ρ in an under-doped ($x = 0.04$) LSCO film oscillates as $\sin(2\varphi)$, breaking the 4-fold rotational symmetry of the crystal structure as shown in Figure 8.12(b). That the anisotropy emerges in superconducting samples may not be a coincidence, and its origin was attributed to electronic nematicity.

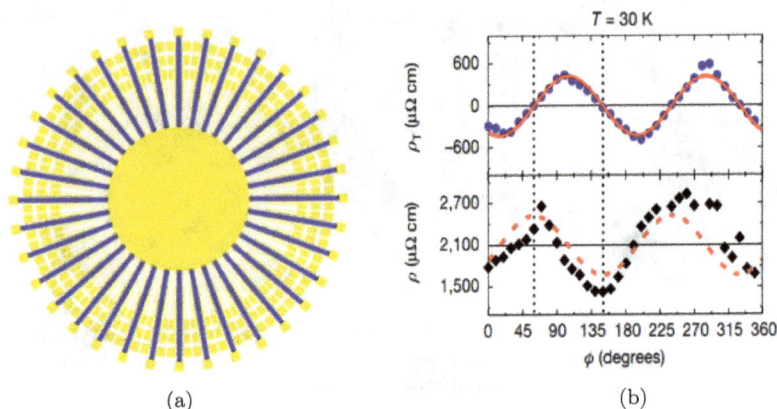

(a)   (b)

Figure 8.12.   (a) The "sunbeam" lithography pattern used for device fabrication contains 36 Hall bars, each with three pairs of transverse contacts. (b) The measured values of $\rho_T(\varphi)$ and $\rho(\varphi)$ at $T = 30$ K.

The spread of such method is still restricted by the difficulties in micro-fabrications. Shan *et al.* developed a scanning point-contact probe compatible with the PPMS. By mounting the sample on a moving stage consisting of close-loop z-axis and x-axis piezo cubes, the Pt–Ir tip can be positioned on the sample surface with an accuracy of sub-micron scale. With this probe, He *et al.* investigated the tunneling spectra of [111]-, [110]-, and [001]-oriented high quality $LiTi_2O_4$ (LTO) thin films and unveiled anisotropic electron–phonon coupling in LTO system [69]. The capability of probing local physical properties of superconductors will allow to rapidly screen the super-conducting combinatorial film without complex micro-fabrication.

8.3.3. *Magnetic properties*

Since probes with beams focused by optical lenses are mostly com-mercially available, here we skip the description of magnetic-optical methods for high-throughput characterizations. Scanning magnetic microscopes, e.g., magnetic force microscope (MFM) [70–72], scan-ning Hall probe microscope (SHPM) [73, 74] and scanning super-conducting quantum interference device (SQUID) [75–77], which possess the spatial resolution necessary for high-throughput magnetic imaging, have played and will continue to play an important role in understanding the nature of superconductivity.

MFM, with spatial resolution of about 100 nm, is generally based on measuring the force between a magnetized tip and the scanned surface. This method shows potential for high-throughput mapping in both static and dynamic magnetic fields [78]. MFM can image topography and even a single quantized vortex in superconducting materials [79].

As a sensitive but non-destructive instrument, SHPM is a promising high-throughput magnetic microscope. It provides quantitative measurements of magnetic field with spatial resolution better than $1\,\mu$m under variable condition. A sub-micron Hall probe can be manufactured and scanned over a sample with the aid of conventional scanning techniques [80]. Similar to MFM, SHPM can also image the distribution of quantized vortices in type-II superconductors [81].

Local magnetic fields can also be imaged by scanning samples with SQUIDs, which utilize the effects of Cooper pair tunneling and superconducting quantum interference [82]. In contrast to the MFM and SHPM, scanning SQUIDs can obtain information about the local susceptibility and superfluid density [83].

In addition to the transport and magnetic properties characterization tools mentioned above, near-field detection techniques like scanning tip microwave near-field microscope (STMNM) have been developed to probe the dynamical electromagnetic properties of superconductors.

Wei *et al.* used a sharpened solid metal instead of an aperture or gap as a point-like evanescent field emitter. Based on this design, they set up a near-field rf/microwave microscope that is capable of imaging surface resistance profiles with a spatial resolution of $\sim$5 μm [84]. Later, Gao *et al.* improved the spatial resolution to the order of 100 nm by introducing a sapphire disk with a center hole comparable to the diameter of the tip wire and a metal coating layer of $1\,\mu$m on the surface as shown in Figure 8.13. Besides, the phase-sensitive technique was applied for faster data acquisition, while quasi-static theoretical model was developed to calculate the relative dielectric constant from the raw data [85].

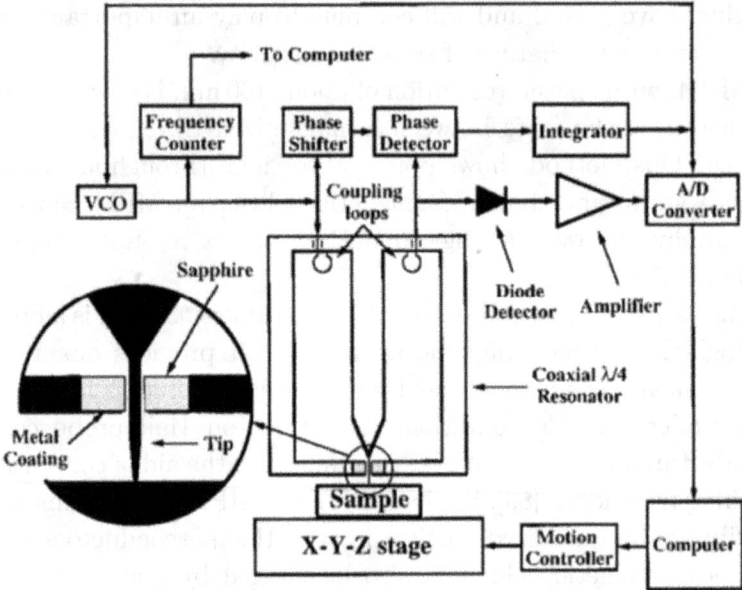

Figure 8.13.   Schematic of the experimental setup for the STMNM with a resonator.

In the same year, Takeuchi *et al.* integrated the STMNM with a low-temperature cryostat and performed the scanning on $YBa_2Cu_3O_{7-\delta}$ film. The high-frequency superconductive response at different positions of the film can be well distinguished [86].

Shen group designed a near-field scanning microwave microscope in a different way. They separated the sensing probe from the excitation electrode to suppress the common-mode signal and used co-planar waveguides to transmit microwave signal [87]. Similarly, their microscope was later upgraded for low-temperature imaging [88].

In addition to composition, structure, electrical transport, and magnetism, methods were developed to image thermal, optical, and mechanical parameters as well [89–94]. From these measurements, some key information of the unconventional superconductivity can be derived.

For example, Otani *et al.* set up an integrated scanning-spring-probe system for high-throughput thermoelectric power-factor

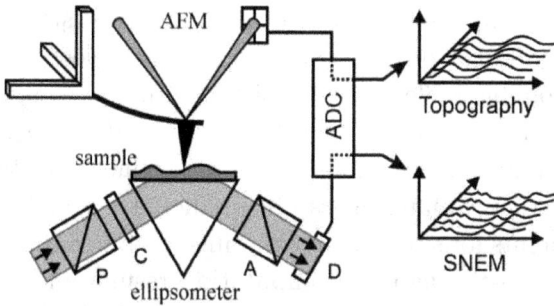

Figure 8.14. Experimental setup of scanning near-field ellipsometric microscope [98].

screening [92]. The thermal transport signals such as Nernst and thermopower are complementary to the electrical transport signals, and indispensable in clarifying issues such as multiband features [95], superconducting fluctuations [96] and phase transitions [97].

An imaging ellipsometer with a lateral resolution in nanometer range was invented by Karageorgiev *et al.* As shown in Figure 8.14, the system combines an ellipsometer with atomic force microscope (AFM). The sample is illuminated from the bottom through a prism in total internal reflection (TIR) and the evanescent field created at the sample surface is scattered by a conventional AFM tip [98, 99]. The intensity of the measured optical signal depends on sample thickness and local optical properties. By fitting the data and making a comparison with an existing database, the local thickness and parameters such as refractive index and absorption factor can be extracted [100, 101]. This sheds light on some interesting topics like negative index induced by surface Josephson-plasma waves in layered superconductors [102].

8.4. High-Throughput Theoretical Modeling for Superconductors

8.4.1. *Introduction to high-throughput computations*

In combinatorial materials studies, the space of possible combinations is so enormous that even high-throughput experiments can rarely or never cover it completely. So the initial candidate systems needs some intuitive, theoretical or past-experience-based selection in order to achieve reasonable scale and rate of predicting, screening and optimizing [103–106]. By using high-throughput *ab-initio* calculations, molecular dynamics simulations, Monte Carlo method, etc., variety of properties for many materials can be predicted at a high rate, serving as starting points for subsequent experiments. Furthermore, the combination of high-throughput computation results and experimental data will facilitate the construction of materials databases. This, in turn, will allow researchers to use materials informatics and artificial intelligence to analyze and mine these databases, and to provide ideas for new materials designs. High-throughput computation is one of the three key links in the Materials Genome Initiative (MGI) [107]. Only when computations are combined with high-throughput experiments and materials databases, the on-demand design of materials can be truly realized.

High-throughput computations have already been employed in some research fields, such as solar water splitters [108, 109], solar photovoltaics [110], topological insulators [111], scintillators [112, 113], CO_2 capture materials [114], piezoelectrics [115], thermoelectrics [116, 117], hydrogen storage materials [118, 119], Li-ion batteries [120–124].

8.4.2. *High-throughput computations for superconductors*

Since the discovery of superconductivity, searching for superconductors with even higher transition temperatures has been a core goal of the field. However, as the number of elements increases, the set of possible compounds grows exponentially. If we further consider the ratio of composed elements, synthesis conditions and other factors, it seems impossible to cover the entire parameter space. In general, many superconductors are sensitive to small changes in composition and synthesis conditions, and a little variation in these parameters may cause significant change in transition temperature. Thus, implementing robust computational tools for materials

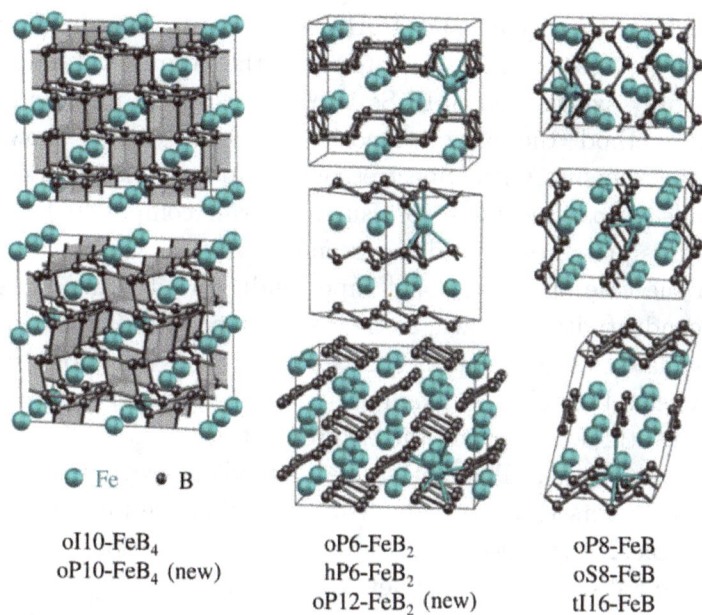

Figure 8.15. Competing B-rich Fe–B phases.

prediction, screening and optimization will make discovery of new superconductors much faster and more efficient.

There have been some successful examples of discovery of superconductors with the help of high-throughput computations. Kolmogorov *et al.* undertook an intensive study of the Fe–B system via *ab-initio* high-throughput evolutionary calculations [125]. The computations for crystal structure and formation energy cover the entire Fe–B system as shown in Figure 8.15. The calculation of electron–phonon coupling predicted an $oP10$-FeB_4 conventional superconductor. Stimulated by this theory study, a high-throughput combinatorial research was subsequently devised and executed, mentioned in previous part [30].

Matsumoto *et al.* employed high-throughput computations for new thermoelectric or superconducting materials [126, 127]. By data-driven materials search, they extracted materials with some specific characteristics (e.g., flat band) as candidates from the inorganic materials database (AtomWork). Further first-principle calculations screened out $SnBi_2Se_4$ and $PbBi_2Te_4$ from these candidates on basis of gap size, density of states near the Fermi level, etc. They successfully synthesized $SnBi_2Se_4$ single crystal via melt and slow-cooling method (the exact composition is $SnBi_{1.64}Se_{3.53}$ showed by Energy Dispersive X-Ray Spectroscopy). Superconducting transition was indeed observed under pressure. A sister compound $PbBi_2Te_4$ also shows pressure-induced superconductivity.

In the case of conventional superconductors, the mechanism of superconductivity is known, so their superconducting properties can be calculated from first principles, at least in theory. However, the mechanisms behind most examples of unconventional superconductivity are still not clear and the electrons in these materials are often strongly correlated. Thus, attempts to predict new unconventional superconductors through deductive calculation are impossible even in principle. Nevertheless, high-throughput computations can at least lower the risks in looking for new superconductors.

High-throughput computations also speed up the construction of database containing basic properties of new materials, which can be combined with machine learning methods to predict new superconductors (see Section 8.5), thus providing a new pathway to superconductivity research [128].

8.5. Machine Learning Methods in Study of Superconductivity

High-throughput experiments often result in large datasets. Since it can be slow and impractical to analyze this data manually, the analysis step has become a common bottleneck for the entire process. Recent developments, however, allow a different approach to data analysis and, even more generally, to exploration of superconductors and the factors that determine the critical temperatures of various materials. Machine learning and statistical methods can use the information amassed in databases covering various measured and calculated materials properties in order to predict macroscopic variables, circumventing to need for theoretical models and complicated *ab-initio* tools.

The difficulties of modeling strongly correlated materials such as cuprate superconductors have for decades inspired the search for novel theoretical approaches. Application of statistical methods in the context of superconductivity began with simple clustering methods in the space of several phenomenological elemental features [129, 130]. Another early work used statistical methods to look for correlations between normal state properties and T_c of the metallic elements in the first six rows of the periodic table [131].

For years these early works remained isolated attempts to use data-driven approaches for study of superconductivity. However, the recent significant increase of available data and the widespread adoption of machine learning algorithms to address variety of research questions changed that. An investigation in 2015 again discovered a clustering property of superconductors using structural and electronic properties data as predictors [132]. A classification model separating superconductors into two groups according to their T_c was developed and showed good performance. As another demonstration of the impact machine learning can have in the search for superconductors, a sequential learning framework designed to discover the material with the highest T_c was evaluated on ∼600 known superconductors [133]. The framework did significantly better than a method based on random guessing.

A work in 2018 built upon many of the lessons from these earlier studies [32]. Whereas previous investigations explored several hundred compounds at most, it considered more than ten thousand different compositions. These were extracted from the SuperCon database, created and maintained by the National Institute for Materials Science in Japan. The database contains relevant information such as T_c for thousands of superconducting materials, collected from journal publications. It is an exhaustive list of all reported superconductors, as well as related non-superconducting compounds, including many closely related materials varying only by small changes in stoichiometry. This information permitted to study the importance of small changes of chemical composition among related compounds.

From SuperCon, a list of approximately 16,400 compounds was extracted, of which 4000 have no T_c reported. Of these, roughly 5700 compounds are cuprates and 1500 are iron-based (about 35% and 9%, respectively), reflecting the significant research efforts invested in these two high-T_c families. The remaining set is a mix of various materials, including conventional phonon-driven superconductors and exotic superconductors such as heavy-fermion compounds and Chevrel phases.

A serious problem is that for most materials SuperCon provides only the chemical composition and T_c. To convert this information into meaningful predictors suitable for machine learning modeling, the Materials Agnostic Platform for Informatics and Exploration (Magpie) was employed [134]. Magpie generates a set of attributes for each material, including elemental property statistics like the mean and the standard deviation of 22 different elemental properties (e.g., period/group on the periodic table, atomic number, atomic radii), as well as attributes such as the average fraction of electrons from the s, p, d, and f valence shells among all elements present.

Incorporating features like lattice type and density of states can lead to significantly more powerful and interpretable models. However, since such information is not generally available in SuperCon, data from the AFLOW Online Repositories [135] was employed. The materials database contains extensively validated calculated properties for the vast majority of compounds in the Inorganic Crystal Structure Data Base (ICSD) [136]. For the SuperCon materials contained within AFLOW (about 600 with finite T_c), a set of 26 additional predictors was calculated, including structural/chemical information such as the lattice type, as well as electronic properties including the density of states near the Fermi level.

First, a classification model designed to separate materials into two distinct groups depending on whether T_c is above or below some predetermined value was developed. This approach resolves the problem of the potentially ambiguous $T_c = 0$ K prescription for materials without reported T_c. The temperature that separates the two groups — T_{sep} — was treated as an adjustable parameter of the model. Compounds with no reported T_c were included in the below-T_{sep} bin. Since setting T_{sep} too low or too high creates strongly imbalanced classes (with most superconductors in one of groups), several different metrics were used to compare the models. For a realistic estimate of the performance of each model, the dataset was randomly split (in 85%:15% proportion) into training and test subsets. The training set was used to fit the model, which was then applied to the test set for benchmarking. A metric calculated on the test set is an unbiased estimate of how the method is expected to perform when applied to a new (but similar) dataset. Ultimately, $T_{sep} = 10$ K was selected; for this T_{sep} the proportion of above-T_{sep} materials is approximately 38% and the accuracy is about 92%, i.e., the model can correctly classify nine out of ten materials, which is much better than random guessing. The recall — quantifying how well above-T_{sep} compounds are correctly labelled (thus an important metric when searching for new superconducting materials) — is even higher.

To create a model predicting the actual T_c values, the same set of predictors was used. One model predicting $\log(T_c)$ was trained and tested on the materials from the entire Supercon dataset materials with $T_c > 10$ K. The model showed good performance (see Figure 8.16). However, the Supercon data contains distinct families of superconductors, including cuprate and iron-based high-temperature superconductors, with all others denoted "low-T_c" for brevity. Three separate models were constructed and trained only on materials from these groups.

Differences in important predictors across the family-specific models reflect the fact that distinct mechanisms are responsible for driving superconductivity in these groups. Low average atomic weight is a necessary (but not sufficient) condition for increasing T_c among the low-T_c group (atomic weight plays a significant role in conventional superconductors through the Debye frequency of phonons). In the case of cuprate superconductors, a similar limit on T_c is based on the average number of unfilled orbitals. Such findings validate the ability of machine learning approaches to discover meaningful patterns that encode true physical phenomena.

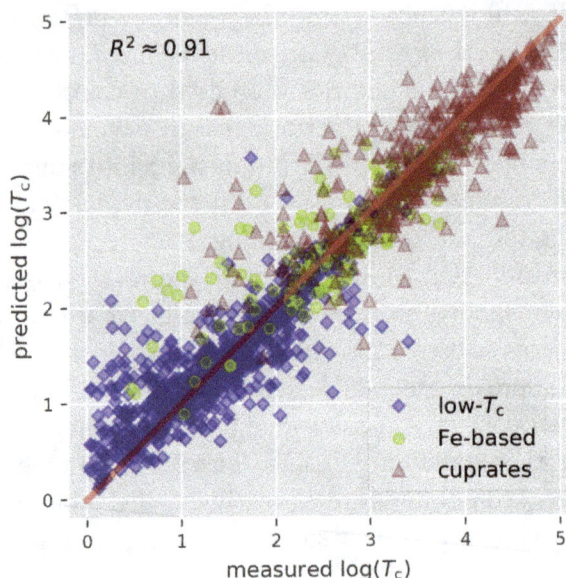

Figure 8.16.   Performance of a regression model for $\log(T_c)$ analogous to the one in [32]. There is a clear correlation between measured and predicted values. Only materials with $T_c > 1$ K were used to train the model.

Finally, the classification and regression models were integrated in a pipeline and employed for a high-throughput screening of the entire ICSD database for candidate superconductors. As a first step, full set of predictors were generated for all compounds in ICSD. Classification model similar to the one described above was constructed, but trained only on materials from SuperCon but not in ICSD. The model was then applied on the ICSD set to create a list of materials with predicted T_c above 10 K. Next in the pipeline, the list was fed into a regression model trained on the entire SuperCon database to predict T_c. Filtering for the materials with $T_c > 20$ K, the list was further reduced to about 2000 compounds.

The vast majority of the compounds identified as candidate superconductors were cuprate materials. There are also some materials clearly related to the iron-based superconductors. The remaining set has 35 members, and is composed of materials that are not obviously connected to any known superconducting family. None of these materials have a predicted T_c in excess of 40 K (no such instances exist in the training dataset). All contain oxygen — also not a surprising result, since the group of known superconductors with $T_c > 20$ K is dominated by oxides.

The list comprises of several distinct groups. Most of the materials are insulators, similar to stoichiometric (and underdoped) cuprates that generally require charge doping and/or pressure to drive these materials into a superconducting state. Especially interesting are the compounds containing heavy metals (such as Au, Ir, Ru), metalloids (Se, Te), and heavier post-transition metals (Bi, Tl), which are or could be pushed into interesting/unstable oxidation states. The most surprising and nonintuitive of the compounds in the list are the silicates and the germanates. Converting these insulators to metals (and possibly superconductors) will require significant charge doping.

Although is not yet clear if any of the predicted compounds are superconducting, the presence of one highly unusual electronic structure feature in many of the candidates is encouraging; almost all of them exhibit one or several flat (or nearly flat) bands just below the energy of the highest occupied electronic state. Such bands lead to a large peak in the density of states and can cause a significant enhancement in T_c. Attempts to synthesize several of these compounds are already underway.

8.6. Conclusion and Perspective

Exploration of the microscopic mechanisms of high-T_c superconductivity and discovery of new unconventional superconductors have always been important topics in condensed matter physics. As materials of interest become more and more complex, the advantages of high-throughput methodology become even more pronounced. In this chapter, we summarized the recent developments in superconductivity-related high-throughput studies. In order to further accelerate the research on superconductors, as well as other functional materials, advances in the following three areas are highly desirable.

8.6.1. *New high-throughput techniques*

While many combinatorial film preparation techniques have been developed, there is a great demand for high-precision characterization instruments, especially ones that can perform advanced spectroscopy measurements such as scanning tunneling microscope and angle resolved photoemission spectroscopy. This means that the combinatorial material library in the form of a film should be measured *in-situ*. Accordingly, combined systems such as combinatorial-film-STM [142] and combinatorial-film-ARPES will be optimal. These will be especially beneficial for the study of superconducting combinatorial libraries. Characterization methods taking advantage of synchrotron sources will become widespread; these techniques allow the study of not only composition and structure of materials, but also of the elementary excitations in condensed matter systems. For instance, resonant inelastic X-ray scattering (RIXS) technique can be used to measure spin excitation [137], charge excitation [138], phase of the order parameter [139], etc. Importantly, since RIXS needs only small sample volumes, thin films can be probed with such techniques. Compared with the neutron scattering technique, RIXS appears more promising in high-throughput research on superconductivity.

8.6.2. *Future high-throughput superconductivity research style*

The high-throughput materials research paradigm is quite different from the traditional approaches. Hanak proposed an integrated materials development workflow about 40 years ago [20]. Potyrailo and Mirsky extended the workflow by adding new elements, such as planning of experiments and data mining as shown [140].

Superconductivity research will certainly benefit from more widespread adoption of this style, but it also has its own specifics. Currently, some characterization methods such as STM, neutron scattering, ARPES, and NMR, commonly used in the field, lag behind the requirements of high-throughput screening. Thus, research on superconductivity, especially on mechanisms of high-temperature superconductivity, has to combine studies of combinatorial libraries with conventional uniform films or bulk crystal samples. This was exemplified in the recent work on the relation between micro-structure and T_c in FeSe superconductor; the evolution of electrical transport properties and crystal lattice parameters was delineated by high-throughput experiments, whereas the electronic band structure was obtained from ARPES on uniform films [51].

8.6.3. *MGI platform*

High-throughput methods with their enormous potential to accelerate materials research have already attracted significant attention. Some countries have started projects and programs in order to support the development of high-throughput computational tools. The US Department of Energy created Scientific Discovery through Advanced Computing (SciDAC) in 2001 to develop new computational methods for tackling some of the most challenging scientific problems. In 2008, the US National Academies created a new discipline, Integrated Computational Materials Engineering, aimed to design materials using a variety of software tools either simultaneously or consecutively. In 2011, the US launched the Materials Project, a core program of the Materials Genome Initiative, that uses high-throughput computing to uncover the properties of all known inorganic materials. "Advanced Concepts in *ab-initio* Simulations of Materials" project, proposed by European Science Foundation, is aimed to develop rapid *ab-initio* calculations, which allow parameter-free calculations of real materials at the atomic level and are applicable to condensed matter systems

However, in order to maximize the efficiency and impact of high-throughput methodology, it is necessary to closely coordinate all the steps in the research cycle. The collected data need to be standardized in order to establish sharable databases. Techniques such as STM, neutron scattering, ARPES, and NMR can take advantage of and contribute to the database according to their own characteristics. Since Materials Genome Initiative (MGI) was launched in the United States in 2011 [141], many similar projects and programs have been initiated. Accelerated Metallurgy (AccMet) project organized by European Union and "MatNavi" database managed by National Institute for Materials Science in Japan are examples of programs established to develop new materials R&D procedures and efficient information sharing mechanisms. Among them, the Center for Materials Genome Initiative (CMGI) in Beijing is the specialized entity organized as an ideal high-throughput platform. This platform contains three substations, focusing on materials computation and data process, high throughput syntheses and fast characterizations, and high throughput technique and development, respectively.

In conclusion, integrating a novel research paradigm with new techniques, facilities, and platforms, high-throughput methodology is becoming more and more prominent in advancing the study of superconducting materials.

Acknowledgments

The authors would like to thank M. Y. Qin, X. Zhang, X. Y. Jiang, G. He, D. Li, R. Z. Zhang, and J. S. Zhang for the help during the manuscript preparation. This work was supported by the Strategic Priority Research Program of Chinese Academy of Sciences (XDB25000000), the National Key Basic Research Program of China (2015CB921000, 2016YFA0300301, 2017YFA0303003, and 2017YFA0302902), the National Natural Science Foundation of China (11674374, 11804378 and 51672307), and the Key Research Program of Frontier Sciences, CAS (QYZDB-SSW-SLH008). The work at the University of Maryland was supported by ONR (N000141512222 and N00014-13-1-0635), and AFOSR (FA 9550-14-10332).

Bibliography

[1] V. L. Ginzburg, and D. A. Kirzhnits, *High-temperature superconductivity* (Springer, 1982).
[2] J. G. Bednorz, and K. A. Muller, *Phys. B Con. Mat.*, **64**, 189 (1986).

[3] S. Adachi, S. Takano, and H. Yamauchi, *J. Alloys Compd.*, **195**, 19 (1993).

[4] W. Korczak, M. Perroux, and P. Strobel, *Physica C*, **193**, 303 (1992).

[5] G. F. Sun, K. W. Wong, B. R. Xu, Y. Xin, and D. F. Lu, *Phys. Lett. A*, **192**, 122 (1994).

[6] Y. Kamihara, T. Watanabe, M. Hirano, and H. Hosono, *J. Am. Chem. Soc.*, **130**, 3296 (2008).

[7] F. C. Hsu, J. Y. Luo, K. W. Yeh, T. K. Chen, T. W. Huang, P. M. Wu, Y. C. Lee, Y. L. Huang, Y. Y. Chu, D. C. Yan, and M. K. Wu, *Proc. Natl. Acad. Sci. USA*, **105**, 14262 (2008).

[8] W. Wu, J. G. Cheng, K. Matsubayashi, P. P. Kong, F. K. Lin, C. Q. Jin, N. L. Wang, Y. Uwatoko, and J. L. Luo, *Nat. Commun.*, **5**, 5508 (2014).

[9] R. Y. Chen, and N. L. Wang, *Rep. Prog. Phys.*, **82**, 012503 (2019).

[10] J. G. Cheng, and J. L. Luo, *J. Phys. Condens. Mat.*, **29**, 383003 (2019).

[11] A. P. Drozdov, M. I. Eremets, I. A. Troyan, V. Ksenofontov, and S. I. Shylin, *Nature*, **525**, 73 (2015).

[12] L. J. Zhang, Y. C. Wang, J. Lv, and Y. M. Ma, *Nat. Rev. Mater.*, **2**, 17005 (2017).

[13] Q. Y. Wang, Z. Li, W. H. Zhang, Z. C. Zhang, J. S. Zhang, W. Li, H. Ding, Y. B. Ou, P. Deng, K. Chang, J. Wen, C. L. Song, K. He, J. F. Jia, S. H. Ji, Y. Y. Wang, L. L. Wang, X. Chen, X. C. Ma, and Q. K. Xue, in *Chinese Phys. Lett.*, **29**, 037402 (2012).

[14] V. Y. Butko, G. Logvenov, N. Bozovic, Z. Radovic, and I. Bozovic, *Adv. Mater.*, **21**, 3644 (2009).

[15] J. T. Ye, Y. J. Zhang, R. Akashi, M. S. Bahramy, R. Arita, and Y. Iwasa, *Science*, **338**, 1193 (2012).

[16] B. Lei, N. Z. Wang, C. Shang, F. B. Meng, L. K. Ma, X. G. Luo, T. Wu, Z. Sun, Y. Wang, Z. Jiang, B. H. Mao, Z. Liu, Y. J. Yu, Y. B. Zhang, and X. H. Chen, *Phys. Rev. B*, **95**, 020503 (2017).

[17] H. Saadaoui, Z. Salman, H. Luetkens, T. Prokscha, A. Suter, W. A. MacFarlane, Y. Jiang, K. Jin, R. L. Greene, E. Morenzoni, and R. F. Kiefl, *Nat. Commun.*, **6**, 6041 (2015).

[18] S. S. Mao, *J. Cryst. Growth*, **379**, 123 (2013).

[19] H. Koinuma, and I. Takeuchi, *Nat. Mater.*, **3**, 429 (2004).

[20] J. J. Hanak, *J. Mater. Sci.*, **5**, 964 (1970).

[21] M. L. Green, I. Takeuchi, and J. R. Hattrick-Simpers, *J. Appl. Phys.*, **113**, 231101 (2013).

[22] M. Iranmanesh, and J. Hulliger, *Prog. Solid State Chem.*, **44**, 123 (2016).

[23] R. Potyrailo, K. Rajan, K. Stoewe, I. Takeuchi, B. Chisholm, and H. Lam, *ACS Comb. Sci.*, **13**, 579 (2011).

[24] Z. H. Barber, and M. G. Blamire, *Mater. Sci. Tech.-Lond.*, **24**, 757 (2008).

[25] S. I. Woo, K. W. Kim, H. Y. Cho, K. S. Oh, M. K. Jeon, N. H. Tarte, T. S. Kim, and A. Mahmood, *Qsar & Combin. Sci.*, **24**, 138 (2005).

[26] X. D. Xiang, *Appl. Surf. Sci.*, **223**, 54 (2004).

[27] J. C. Zhao, *Prog. Mater. Sci.*, **51**, 557 (2006).

[28] J. J. de Pablo, N. E. Jackson, M. A. Webb, L.-Q. Chen, J. E. Moore, D. Morgan, R. Jacobs, T. Pollock, D. G. Schlom, E. S. Toberer, J. Analytis, I. Dabo, D. M. DeLongchamp, G. A. Fiete, G. M. Grason, G. Hautier, Y. Mo, K. Rajan, E. J. Reed, E. Rodriguez, V. Stevanovic, J. Suntivich, K. Thornton, and J.-C. Zhao, *NPJ Comput. Mater.*, **5**, 41 (2019).

[29] X. D. Xiang, X. Sun, G. Briceno, Y. Lou, K. A. Wang, H. Chang, W. G. Wallace-Freedman, S. W. Chen, and P. G. Schultz, *Science*, **268**, 1738 (1995).

[30] K. Jin, R. Suchoski, S. Fackler, Y. Zhang, X. Q. Pan, R. L. Greene, and I. Takeuchi, *Apl Mater.*, **1**, 042101 (2013).

[31] J. Wu, A. T. Bollinger, Y. Sun, and I. Bozovic, *Proc. Natl. Acad. Sci. USA*, **113**, 4284 (2016).

[32] V. Stanev, C. Oses, A. G. Kusne, E. Rodriguez, J. Paglione, S. Curtarolo, and I. Takeuchi, *NPJ Comput. Mater.*, **4**, 29 (2018).

[33] H. Koinuma, H. N. Aiyer, and Y. Matsumoto, *Sci. Technol. Adv. Mat.*, **1**, 1 (2000).

[34] K. Kennedy, T. Stefansk, G. Davy, V. F. Zackay, and E. R. Parker, *J. Appl. Phys.*, **36**, 3808 (1965).

[35] S. Guerin, and B. E. Hayden, *J. Comb. Chem.*, **8**, 66 (2006).

[36] J. R. Dahn, S. Trussler, T. D. Hatchard, A. Bonakdarpour, J. R. Mueller-Neuhaus, K. C. Hewitt, and M. Fleischauer, *Chem. Mater.*, **14**, 3519 (2002).

[37] M. Saadat, A. E. George, and K. C. Hewitt, *Physica C*, **470**, S59 (2010).

[38] H. Koinuma, *Solid State Ionics*, **108**, 1 (1998).

[39] X. D. Xiang, *Biotechnol. Bioeng.*, **61**, 227 (1999).

[40] X. D. Xiang, *Annu. Rev. Mater. Sci.*, **29**, 149 (1999).

[41] X. D. Xiang, *Mat. Sci. Eng. B-Solid*, **56**, 246 (1998).

[42] J. Wang, and J. R. G. Evans, *J. Comb. Chem.*, **7**, 665 (2005).

[43] I. Van Driessche, J. Feys, S. C. Hopkins, P. Lommens, X. Granados, B. A. Glowacki, S. Ricart, B. Holzapfel, M. Vilardell, A. Kirchner, and M. Backer, *Supercond. Sci. Tech.*, **25**, 065017 (2012).

[44] L. Chen, J. Bao, and C. Gao, *J. Comb. Chem.*, **6**, 699 (2004).

[45] E. Reddington, A. Sapienza, B. Gurau, R. Viswanathan, S. Sarangapani, E. S. Smotkin, and T. E. Mallouk, *Science*, **280**, 1735 (1998).

[46] Y. K. Yoo, and X. D. Xiang, *J. Phys. Condens. Mat.*, **14**, R49 (2002).

[47] H. S. Yu, J. Yuan, B. Y. Zhu, and K. Jin, *Sci. China Phys. Mech. Astron.*, **60**, 087421 (2017).

[48] T. Fukumura, M. Ohtani, M. Kawasaki, Y. Okimoto, T. Kageyama, T. Koida, T. Hasegawa, Y. Tokura, and H. Koinuma, *Appl. Phys. Lett.*, **77**, 3426 (2000).

[49] T. M. McQueen, Q. Huang, V. Ksenofontov, C. Felser, Q. Xu, H. Zandbergen, Y. S. Hor, J. Allred, A. J. Williams, D. Qu, J. Checkelsky, N. P. Ong, and R. J. Cava, *Phys. Rev. B*, **79**, 014522 (2009).

[50] T. Ohnishi, M. Lippmaa, T. Yamamoto, S. Meguro, and H. Koinuma, *Appl. Phys. Lett.*, **87**, 241919 (2005).

[51] Z. P. Feng, J. Yuan, J. Li, X. X. Wu, W. Hu, B. Shen, M. Y. Qin, L. Zhao, B. Y. Zhu, H. B. Wang, M. Liu, G. M. Zhang, J. P. Hu, H. X. Yang, J. Q. Li, X. L. Dong, F. Zhou, X. J. Zhou, F. V. Kusmartsev, I. Takeuchi, Z. X. Zhao, and K. Jin, arXiv: 1807.01273 [cond-mat.supr-con].

[52] J. Wu, A. T. Bollinger, Y. J. Sun, and I. Bozovic, *Proc. Natl. Acad. Sci. USA*, **113**, 4284 (2016).

[53] J. Wu and I. Bozovic, *Apl Mater.*, **3**, 062401 (2015).

[54] T. Koida, D. Komiyama, H. Koinuma, M. Ohtani, M. Lippmaa, and M. Kawasaki, *Appl. Phys. Lett.*, **80**, 565 (2002).

[55] T. Chikyow, P. Ahmet, K. Nakajima, T. Koida, M. Takakura, M. Yoshimoto, and H. Koinuma, *Appl. Surf. Sci.*, **189**, 284 (2002).

[56] M. Naito, and M. Hepp, *Jpn J. Appl. Phys. 2*, **39**, L485 (2000).

[57] A. Sawa, M. Kawasaki, H. Takagi, and Y. Tokura, *Phys. Rev. B*, **66**, 014531 (2002).

[58] H. Rarback, D. Shu, S. C. Feng, H. Ade, J. Kirz, I. Mcnulty, D. P. Kern, T. H. P. Chang, Y. Vladimirsky, N. Iskander, D. Attwood, K. Mcquaid, and S. Rothman, *Rev. Sci. Instrum.*, **59**, 52 (1988).

[59] M. Ohtani, T. Fukumura, M. Kawasaki, K. Omote, T. Kikuchi, J. Harada, A. Ohtomo, M. Lippmaa, T. Ohnishi, D. Komiyama, R. Takahashi, Y. Matsumoto, and H. Koinuma, *Appl. Phys. Lett.*, **79**, 3594 (2001).

[60] Z. Luo, B. Geng, J. Bao, C. Liu, W. Liu, C. Gao, Z. Liu, and X. Ding, *Rev. Sci. Instrum.*, **76**, 095105 (2005).

[61] C. A. MacDonald, *Annu. Rev. Mater. Res.*, **47**, 115 (2017).

[62] E. D. Isaacs, M. Marcus, G. Aeppli, X. D. Xiang, X. D. Sun, P. Schultz, H. K. Kao, G. S. Cargill, and R. Haushalter, *Appl. Phys. Lett.*, **73**, 1820 (1998).

[63] J. M. Gregoire, D. G. Van Campen, C. E. Miller, R. J. Jones, S. K. Suram, and A. Mehta, *J. Synchrotron Radiat.*, **21**, 1262 (2014).

[64] H. Xing, B. Zhao, Y. Wang, X. Zhang, Y. Ren, N. Yan, T. Gao, J. Li, L. Zhang, and H. Wang, *ACS Comb. Sci.*, **20**, 127 (2018).

[65] S. K. Suram, S. W. Fackler, L. Zhou, A. T. N'Diaye, W. S. Drisdell, J. Yano, and J. M. Gregoire, *ACS Comb. Sci.*, **20**, 26 (2018).

[66] K. C. Hewitt, P. A. Casey, R. J. Sanderson, M. A. White, and R. Sun, *Rev. Sci. Instrum.*, **76**, 093906 (2005).

[67] J. Wu, O. Pelleg, G. Logvenov, A. T. Bollinger, Y. J. Sun, G. S. Boebinger, M. Vanevic, Z. Radovic, and I. Bozovic, *Nat. Mater.*, **12**, 877 (2013).

[68] J. Wu, A. T. Bollinger, X. He, and I. Bozovic, *Nature*, **547**, 432 (2017).

[69] G. He, Y. L. Jia, X. Y. Hou, Z. X. Wei, H. D. Xie, Z. Z. Yang, J. N. Shi, J. Yuan, L. Shan, B. Y. Zhu, H. Li, L. Gu, K. Liu, T. Xiang, and K. Jin, *Phys. Rev. B*, **95**, 054510 (2017).

[70] C. Bruder, *Physica C*, **185**, 1671 (1991).

[71] H. J. Reittu, and R. Laiho, *Supercond. Sci. Tech.*, **5**, 448 (1992).

[72] A. P. Volodin, and M. V. Marchevsky, *Ultramicroscopy*, **42**, 757 (1992).

[73] A. M. Chang, H. D. Hallen, L. Harriott, H. F. Hess, H. L. Kao, J. Kwo, R. E. Miller, R. Wolfe, J. Vanderziel, and T. Y. Chang, *Appl. Phys. Lett.*, **61**, 1974 (1992).

[74] A. M. Chang, H. D. Hallen, H. F. Hess, H. L. Kao, J. Kwo, A. Sudbo, and T. Y. Chang, *Europhys Lett.*, **20**, 645 (1992).

[75] Y. Matsumoto, M. Murakami, T. Shono, T. Hasegawa, T. Fukumura, M. Kawasaki, P. Ahmet, T. Chikyow, S. Koshihara, and H. Koinuma, *Science*, **291**, 854 (2001).

[76] C. P. Foley, and H. Hilgenkamp, *Supercond. Sci. Tech.*, **22**, 064001 (2009).

[77] S. Bechstein, C. Kohn, D. Drung, J. H. Storm, O. Kieler, V. Morosh, and T. Schurig, *Supercond. Sci. Tech.*, **30**, 034007 (2017).

[78] Y. Martin, and H. K. Wickramasinghe, *Appl. Phys. Lett.*, **50**, 1455 (1987).

[79] A. Volodin, K. Temst, Y. Bruynseraede, C. Van Haesendonck, M. I. Montero, I. K. Schuller, B. Dam, J. M. Huijbregtse, and R. Griessen, *Physica C: Superconductivity*, **369**, 165 (2002).

[80] A. Oral, S. J. Bending, and M. Henini, *Appl. Phys. Lett.*, **69**, 1324 (1996).

[81] H. Bluhm, S. E. Sebastian, J. W. Guikema, I. R. Fisher, and K. A. Moler, *Phys. Rev. B*, **73**, 014514 (2006).

[82] N. C. Koshnick, M. E. Huber, J. A. Bert, C. W. Hicks, J. Large, H. Edwards, and K. A. Moler, *Appl. Phys. Lett.*, **93**, 243101 (2008).

[83] B. Kalisky, J. R. Kirtley, J. G. Analytis, J.-H. Chu, A. Vailionis, I. R. Fisher, and K. A. Moler, *Phys. Rev. B*, **81**, 184513 (2010).

[84] T. Wei, X. D. Xiang, W. G. WallaceFreedman, and P. G. Schultz, *Appl. Phys. Lett.*, **68**, 3506 (1996).

[85] C. Gao, T. Wei, F. Duewer, Y. L. Lu, and X. D. Xiang, *Appl. Phys. Lett.*, **71**, 1872 (1997).

[86] I. Takeuchi, T. Wei, F. Duewer, Y. K. Yoo, X. D. Xiang, V. Talyansky, S. P. Pai, G. J. Chen, and T. Venkatesan, *Appl. Phys. Lett.*, **71**, 2026 (1997).

[87] K. Lai, M. B. Ji, N. Leindecker, M. A. Kelly, and Z. X. Shen, *Rev. Sci. Instrum.*, **78**, 063702 (2007).

[88] W. Kundhikanjana, K. J. Lai, M. A. Kelly, and Z. X. Shen, *Rev. Sci. Instrum.*, **82**, 033705 (2011).

[89] R. A. Potyrailo, and I. Takeuchi, *Meas. Sci. Technol.*, **16**, 1 (2005).

[90] W. C. Oliver, and G. M. Pharr, *J. Mater. Res.*, **7**, 1564 (1992).

[91] J. S. Wang, Y. Yoo, C. Gao, I. Takeuchi, X. D. Sun, H. Y. Chang, X. D. Xiang, and P. G. Schultz, *Science*, **279**, 1712 (1998).

[92] M. Otani, N. D. Lowhorn, P. K. Schenck, W. Wong-Ng, M. L. Green, K. Itaka, and H. Koinuma, *Appl. Phys. Lett.*, **91**, 132102 (2007).

[93] Y. G. Yan, J. Martin, W. Wong-Ng, M. Green, and X. F. Tang, *Rev. Sci. Instrum.*, **84**, 115110 (2013).

[94] G. R. Winkler, and J. R. Winkler, *Rev. Sci. Instrum.*, **82**, 114101 (2011).

[95] W. Jiang, S. N. Mao, X. X. Xi, X. G. Jiang, J. L. Peng, T. Venkatesan, C. J. Lobb, and R. L. Greene, *Phys. Rev. Lett.*, **73**, 1291 (1994).

[96] A. Pourret, P. Spathis, H. Aubin, and K. Behnia, *New J. Phys.*, **11**, 055071 (2009).

[97] O. Cyr-Choiniere, R. Daou, F. Laliberte, D. LeBoeuf, N. Doiron-Leyraud, J. Chang, J. Q. Yan, J. G. Cheng, J. S. Zhou, J. B. Goodenough, S. Pyon, T. Takayama, H. Takagi, Y. Tanaka, and L. Taillefer, *Nature*, **458**, 743 (2009).

[98] P. Karageorgiev, H. Orendi, B. Stiller, and L. Brehmer, *Appl. Phys. Lett.*, **79**, 1730 (2001).

[99] D. Tranchida, J. Diaz, P. Schon, H. Schonherr, and G. J. Vancso, *Nanoscale*, **3**, 233 (2011).

[100] M. L. Zhao, J. Lian, H. S. Yu, K. Jin, L. P. Xu, Z. G. Hu, X. L. Yang, and S. S. Kang, *Appl. Surf. Sci.*, **421**, 611 (2017).

[101] M. L. Zhao, J. Lian, Z. Z. Sun, W. F. Zhang, M. M. Li, Y. Wang, H. S. Yu, K. Jin, and X. Y. Hu, *Opt. Mater. Express*, **5**, 2047 (2015).

[102] V. A. Golick, D. V. Kadygrob, V. A. Yampol'skii, A. L. Rakhmanov, B. A. Ivanov, and F. Nori, *Phys. Rev. Lett.*, **104**, 187003 (2010).

[103] R. Gomez-Bombarelli, J. Aguilera-Iparraguirre, T. D. Hirzel, D. Duvenaud, D. Maclaurin, M. A. Blood-Forsythe, H. S. Chae, M. Einzinger, D. G. Ha, T. Wu, G. Markopoulos, S. Jeon, H. Kang, H. Miyazaki, M. Numata, S. Kim, W. L. Huang, S. I. Hong, M. Baldo, R. P. Adams, and A. Aspuru-Guzik, *Nat. Mater.*, **15**, 1120 (2016).

[104] R. V. Chepulskii, and S. Curtarolo, *Phys. Rev. B*, **79**, 134203 (2009).

[105] X. Jiang, and J. J. Zhao, *Rsc. Adv.*, **5**, 48012 (2015).

[106] G. A. Landrum, and H. Genin, *Mater. Res. Soc. Symp. P.*, **700**, 233 (2002).

[107] *Materials Genome Initiative for Global Competitiveness* (USA National Science and Technology Council, 2011).

[108] I. E. Castelli, D. D. Landis, K. S. Thygesen, S. Dahl, I. Chorkendorff, T. F. Jaramillo, and K. W. Jacobsen, *Energ. Environ. Sci.*, **5**, 9034 (2012).

[109] I. E. Castelli, T. Olsen, S. Datta, D. D. Landis, S. Dahl, K. S. Thygesen, and K. W. Jacobsen, *Energ. Environ. Sci.*, **5**, 5814 (2012).

[110] L. P. Yu, and A. Zunger, *Phys. Rev. Lett.*, **108**, 068701 (2012).

[111] K. S. Yang, W. Setyawan, S. D. Wang, M. B. Nardelli, and S. Curtarolo, *Nat. Mater.*, **11**, 614 (2012).

[112] C. Ortiz, O. Eriksson, and M. Klintenberg, *Comp. Mater. Sci.*, **44**, 1042 (2009).

[113] W. Setyawan, R. M. Gaume, S. Lam, R. S. Feigelson, and S. Curtarolo, *ACS Comb. Sci.*, **13**, 382 (2011).

[114] L. C. Lin, A. H. Berger, R. L. Martin, J. Kim, J. A. Swisher, K. Jariwala, C. H. Rycroft, A. S. Bhown, M. W. Deem, M. Haranczyk, and B. Smit, *Nat. Mater.*, **11**, 633 (2012).

[115] R. Armiento, B. Kozinsky, M. Fornari, and G. Ceder, *Phys. Rev. B*, **84**, 014103 (2011).

[116] S. Curtarolo, W. Setyawan, S. D. Wang, J. K. Xue, K. S. Yang, R. H. Taylor, L. J. Nelson, G. L. W. Hart, S. Sanvito, M. Buongiorno-Nardelli, N. Mingo, and O. Levy, *Comp. Mater. Sci.*, **58**, 227 (2012).

[117] S. D. Wang, Z. Wang, W. Setyawan, N. Mingo, and S. Curtarolo, *Phys. Rev. X*, **1**, 021012 (2011).

[118] S. V. Alapati, J. K. Johnson, and D. S. Sholl, *J. Phys. Chem. B*, **110**, 8769 (2006).

[119] J. Lu, Z. Z. G. Fang, Y. J. Choi, and H. Y. Sohn, *J. Phys. Chem. C*, **111**, 12129 (2007).

[120] H. L. Chen, G. Hautier, A. Jain, C. Moore, B. Kang, R. Doe, L. J. Wu, Y. M. Zhu, Y. Z. Tang, and G. Ceder, *Chem. Mater.*, **24**, 2009 (2012).

[121] J. C. Kim, C. J. Moore, B. Kang, G. Hautier, A. Jain, and G. Ceder, *J. Electrochem. Soc.*, **158**, A309 (2011).

[122] G. Hautier, A. Jain, H. L. Chen, C. Moore, S. P. Ong, and G. Ceder, *J. Mater. Chem.*, **21**, 17147 (2011).

[123] A. Jain, G. Hautier, C. Moore, B. Kang, J. Lee, H. L. Chen, N. Twu, and G. Ceder, *J. Electrochem. Soc.*, **159**, A622 (2012).

[124] H. L. Chen, G. Hautier, and G. Ceder, *J. Am. Chem. Soc.*, **134**, 19619 (2012).

[125] A. N. Kolmogorov, S. Shah, E. R. Margine, A. F. Bialon, T. Hammerschmidt, and R. Drautz, *Phys. Rev. Lett.*, **105**, 217003 (2010).

[126] R. Matsumoto, Z. F. Hou, H. Hara, S. Adachi, H. Takeya, T. Irifune, K. Terakura, and Y. Takano, *Appl. Phys. Express*, **11**, 093101 (2018).

[127] R. Matsumoto, Z. F. Hou, M. Nagao, S. Adachi, H. Hara, H. Tanaka, K. Nakamura, R. Murakami, S. Yamamoto, H. Takeya, T. Lrifune, K. Terakura, and Y. Takano, *Sci. Technol. Adv. Mat.*, **19**, 909 (2018).

[128] R. M. Geilhufe, S. S. Borysov, D. Kalpakchi, and A. V. Balatsky, *Phy. Rev. Mater.*, **2**, 024802 (2018).

[129] K. M. Rabe, J. C. Phillips, P. Villars, and I. D. Brown, *Phys. Rev. B*, **45**, 7650 (1992).

[130] P. Villars, and J. C. Phillips, *Phys. Rev. B*, **37**, 2345 (1988).

[131] J. E. Hirsch, *Phys. Rev. B*, **55**, 9007 (1997).

[132] O. Isayev, D. Fourches, E. N. Muratov, C. Oses, K. Rasch, A. Tropsha, and S. Curtarolo, *Chem. Mater.*, **27**, 735 (2015).

[133] J. Ling, M. Hutchinson, E. Antono, S. Paradiso, and B. Meredig, *Integr. Mater. Manuf. Innov.*, **6**, 207 (2017).

[134] L. Ward, A. Agrawal, A. Choudhary, and C. Wolverton, *NPJ Comput. Mater.*, **2**, 16028 (2016).

[135] S. Curtarolo, W. Setyawan, G. L. W. Hart, M. Jahnatek, R. V. Chepulskii, R. H. Taylor, S. D. Wanga, J. K. Xue, K. S. Yang, O. Levy, M. J. Mehl, H. T. Stokes, D. O. Demchenko, and D. Morgan, *Comp. Mater. Sci.*, **58**, 218 (2012).

[136] G. Bergerhoff, R. Hundt, R. Sievers, and I. D. Brown, *J. Chem. Inf. Comp. Sci.*, **23**, 66 (1983).

[137] M. Minola, Y. Lu, Y. Y. Peng, G. Dellea, H. Gretarsson, M. W. Haverkort, Y. Ding, X. Sun, X. J. Zhou, D. C. Peets, L. Chauviere, P. Dosanjh, D. A. Bonn, R. Liang, A. Damascelli, M. Dantz, X. Lu, T. Schmitt, L. Braicovich, G. Ghiringhelli, B. Keimer, and M. Le Tacon, *Phys. Rev. Lett.*, **119**, 097001 (2017).

[138] C. C. Chen, B. Moritz, F. Vernay, J. N. Hancock, S. Johnston, C. J. Jia, G. Chabot-Couture, M. Greven, I. Elfimov, G. A. Sawatzky, and T. P. Devereaux, *Phys. Rev. Lett.*, **105**, 177401 (2010).

[139] P. Marra, S. Sykora, K. Wohlfeld, and J. van den Brink, *Phys. Rev. Lett.*, **110**, 117005 (2013).

[140] R. A. Potyrailo, and V. M. Mirsky, *Chem. Rev.*, **108**, 770 (2008).

[141] A. White, *Mrs Bull.* **37**, 715 (2012).

[142] G. He, Z. X. Wei, Z. P. Feng, X. D. Yu, B. Y. Zhu, L. Liu, K. Jin, J. Yuan and Q. Huan, *Rev. Sci. Instrum.* **91**, 013904 (2020).

Chapter 9

Engineering Epitaxial Superconductor–Semiconductor Heterostructures Using Molecular Beam Epitaxy

Kaushini S. Wickramasinghe* and Javad Shabani[†]

*Center for Quantum Phenomena, Department of Physics,
New York University, NY 10003, USA*
*kskwick@gmail.com
[†]js10080@nyu.edu

9.1. Introduction

Molecular beam epitaxy (MBE) is a sophisticated thin-film deposition system where it allows atomic layer-by-layer crystal growth in ultra-high vacuum environment. MBE has been widely used in semiconductor technology from transistor to optoelectronic devices. The challenge to get the highest purity materials in each layer and have the exquisite control over interface when changing the composition or compound distinguishes MBE from other deposition techniques.

Recently, topological states in solid-state systems have revealed new directions in condensed matter physics with potential applications in topological quantum information. It is realized that if we could find a perfect interface between conventional superconductors (e.g., Al, Nb) and semiconductors with strong spin–orbit coupling (SOC) (e.g., InAs, InSb), we would be able to engineer topological superconductivity that hosts Majorana bound states at the boundary

of the device. The interface allows electrons to inherit superconducting correlations from the *s*-wave superconductor while retaining
large SOC and large *g*-factor from the semiconductor. This balance
depends on how the electron wave function resides in both materials.
First experiments on nanowires grown by chemical vapor deposition
(CVD) revealed zero-bias peak in conductance signature of Majorana
bound states [1]. It was soon realized that MBE is a unique tool to
achieve nearly perfect interface to host topological superconductivity
[2]. Moreover, in order to eventually move beyond demonstrations of
braiding, to larger-scale Majorana networks, it is likely that a top-
down patterning approach will be needed. MBE growth of large-area
two-dimensional (2D) S–Sm systems can form the basis for such an
approach. We discuss the basics of MBE growth and discuss devices
based on epitaxial semiconductors and superconductors.

9.2. Molecular Beam Epitaxy

Molecular beam epitaxy (MBE) was first developed in the 1960s and
1970s to grow layers of single crystals made of semiconductor materials with a burst of computation and communication technologies.
MBE proved as a prominent method to realize ultra-thin layers of
precise thickness. The concentration of unintentional impurities is
very low due to an ultra-high-vacuum (UHV) growth environment
and high-purity source materials. Nowadays, MBE can produce
complex structures with exotic properties like complex oxides, topological insulators in addition to standard III–V heterostructures in
large scale. In MBE growth, molecular beam produced by thermally
evaporating an elemental source which is then deposited onto a
heated crystalline substrate to form the crystalline material. To grow
a layer with high-purity we use extremely pure material sources,
typically purity of 99.9999999% for group III materials (In, Ga and
Al) and 99.99999% purity for group V materials (As and Sb) are
needed. Despite many 9s in the purity of these materials, they still
lack the purity needed. For example, in GaAs growth the background
impurity is below $10^{13} \mathrm{cm}^{-3}$ which requires many more 9s but luckily
not all impurities ionize to contribute as a dopant and affect the band

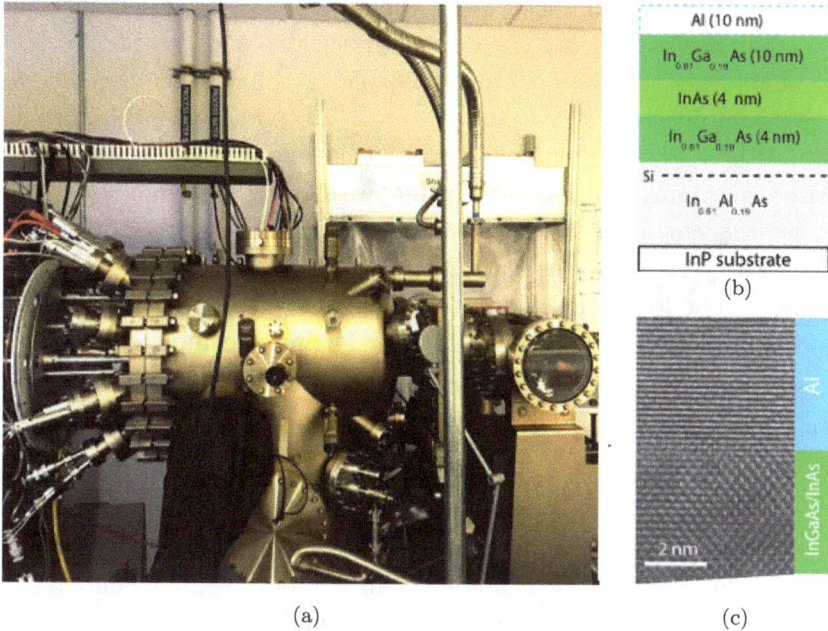

Figure 9.1. (a) The standalone modified Intevac modular Gen II system. The photograph of the side view of the growth chamber. (b) A schematic of the standard Al–InAs structure. (c) High-resolution transmission microscopy image showing the sharp and uniform interface between Al and the InGaAs layer.

diagram of the material. An operational MBE system is typically operated under ultra-high vacuum conditions of $\sim 10^{-10} - 10^{-11}$ Torr.

We use a standalone modified Intevac modular Gen II system for growth of S–Sm structures. Figure 9.1(a) shows a photograph of the MBE systems installed in our laboratory at New York University. This consists of three chambers: an entry–exit or load–lock chamber, a buffer chamber and a growth chamber. All three chambers are isolated from each other via a gate valve. The wafers are transferred between the chambers using a trolley. There are three critical chambers in an MBE system.

9.2.1. *Entry–exit chamber (load–lock)*

The entry–exit chamber is used to load the substrates into the system and bring the grown wafers out of the system without exposing the other chambers to atmospheric pressure. We use dry nitrogen gas to vent this chamber and let the nitrogen gas flow throughout the time the chamber is open to prevent gas molecules outside the chamber from entering the system. Once the wafer exchange is completed, we use a combination of two pumps, turbo pump and a scroll pump, to pump the entry–exit chamber from atmosphere to $\sim 5 \times 10^{-5}$ Torr. After the chamber reaches this value, we switch to a small ion pump. When the vacuum is down to $\sim 5 \times 10^{-7}$ Torr, the entry–exit chamber is baked at 150°C for at least 2 hr to outgas the substrates and the holders. The ion gauge mounted in the entry–exit chamber reads a vacuum of the order $\sim 10^{-9}$ Torr after baking is completed.

9.2.2. Buffer chamber

This is the intermediate chamber connected directly to the growth chamber and the entry–exit chamber. It is maintained at a vacuum of $\sim 10^{-10}$ Torr by an ion pump. The buffer chamber serves the purpose of isolating the growth chamber from the entry–exit chamber and thus helps to maintain the high purity environment in the growth chamber. This chamber also provides a space to temporarily store wafers before and after growth.

9.2.3. Growth chamber

A schematic photograph of the MBE growth chamber is shown in Figure 9.1(a). The major components of the growth chamber include liquid nitrogen-cooled cryoshrouds enclosing much of the volume around the substrate holder, effusion cells to produce the atomic or molecular beams, shutters to mechanically switch the beam fluxes on and off, rotatable substrate holder with the heater, ion gauges to monitor the vacuum, a beam flux measurement gauge, RHEED setup to monitor the *in-situ* growth, a pyrometer to measure the substrate temperature and a mass spectrometer to analyze residual gases and for leak detection.

The feasibility of growing high-quality epilayers in the chamber is due to the removing any elements expect the ones needed for thin film in the growth environment. To achieve this, the following conditions are necessary: first, use of high purity source materials in the effusion cells and use of a strong vacuum pumping system, a combination of a cryopump, ion pump or turbo pump. Since UHV is a result of the steady state between the rate of gas evolution and the pumping rate, these are not sufficient conditions. Hence, two more conditions are applied in constructing an MBE system. One of the conditions is making MBE components with materials that outgas negligibly at higher temperatures, like molybdenum and tantalum. The crucibles are made of high purity pyrolytic boron nitride, which has low outgassing up to a temperature of ∼1500°C. The other one is minimizing the molecules and atoms escaping from the walls of the chamber and heated components of the apparatus reaching the substrate. This is achieved by the liquid nitrogen-cooled cryoshrouds enclosing the volume around the substrate holder. For more details, we refer the readers to [3].

9.3. Hybrid Structure of Epitaxial Surface InAs 2DEG

Semiconductor-based devices have the potential to provide the rapid control and tunability needed for many high-speed circuits. They are widely used in all cell phones, processors, sensors and the communication. In addition, they offer low-power consumption and promising prospects for scalability, make them particularly attractive for industry. Superconductors from the other hand host dissipationless transport and are extremely sensitive to magnetic flux. Superconducting quantum interference device (SQUID) is a famous example of such devices fabricated based on superconducting loops containing Josephson junctions (JJs). SQUIDs are sensitive enough to measure magnetic fields as small as one femtotesla. There are many advantages if semiconductors can be integrated with superconducting components. Devices based on both materials could explore the macroscopic superconducting properties of the superconducting leads with the microscopic degrees of freedom of the semiconductor. In semiconductors, density and conductivity can be tuned using gate voltages, which allows tuning superconducting properties such as supercurrent. Applications vary from ultra-low power Josephson-field effect transistor circuits to tunable superconducting qubits, the so-called gatemon where the Josephson energy can be tuned *in-situ* with an applied electric field.

Recently, realizing transparent contact in superconductor–semiconductor (S–Sm) systems has become the focus of renewed theoretical and experimental attention partly because of the potential applications in spintronics, topological superconductivity [4, 5], and superconducting quantum computation [6–8]. Generally, materials considered for S–Sm systems, such as GaAs two-dimensional electron gas (2DEG) [9] contacted with either aluminum or niobium-based superconductors, have had high quality 2DEG's but suffered from imperfect interfaces due to the presence of a Schottky barrier. Narrow bandgap materials such as InAs and InSb have been studied due to the potential for transparent metallic interfaces [10], first using 2DEGs [11–13] and more recently using nanowires [14]. Recently, it has been shown that epitaxial contacts to nanowires and near surface quantum wells can improve the proximity effect in Josephson junctions [15, 16].

We first focus on the semiconductor hosting a heterostructures containing InAs layers that are hypothesized to be suitable systems for spintronics applications [17]. Spintronics favors materials like InAs with strong spin–orbit coupling (SOC) and large g-factor. Recently, two-dimensional electron systems (2DESs) conned to surface InAs layers have become the focus of renewed theoretical and experimental attention partly because of their potential applications in superconducting quantum computation [6, 7] and realization of topological states of matter [4, 18]. A key feature to these applications is controlled proximity effect with superconductors [19, 20]. InAs is well known to be among a few materials which can interface well with metals and superconductors forming ohmic contacts unlike well-known Si and GaAs materials where the interface forms a Schottky barrier. In the past few decades, studies have been focused on high mobility 2DES where the quantum well is placed tens or hundreds of nanometers from the surface [11–13, 21–26]. To make proximity devices in these heterostructures, contacts have to be made after etching the top barriers with *in-situ* ion cleaning. The fabrication difficulties have limited fabricating and testing of more complicated devices [27] since the interface quality/characteristics are proven difficult to finely tune. With the renewed interest in the field of topological superconductivity, near surface quantum well structures with epitaxial superconducting contacts were developed [16, 28, 29].

One question arises is that how well we can control the 2DES at the surface. There are a few key characteristics that we need to consider for designing a surface 2DES to be able to study the proximity effect in Josephson devices. One important factor in controlling the strength of proximity effect by composition and the thickness of layers is coupling 2DES with surface. Second characteristic is the high carrier mobility that should be achieved for an optimized insertion layer. Finally, the third feature is having a gate tunable spin–orbit coupling.

The near surface quantum wells are grown on semi-insulating InP (100) substrates in a modified Gen II MBE system. The step graded buffer layer, $In_xAl_{1-x}As$, is grown at low temperature to minimize dislocations forming due to the lattice mismatch between the active region and the InP substrate [30–32]. The quantum well consists of a 4 nm layer of InAs grown on a 4 nm layer $In_{0.81}Ga_{0.19}As$. This layer helps to recover the InAlAs buffer roughness. We do not observe any improvement in transport mobility when the thickness of this layer is thicker than 4 nm consistent with earlier studies (30). A delta-doped Si layer of $\sim 7.5 \times 10^{11}$ cm^{-2} is place 6 nm below the lower 4 nm $In_{0.81}Ga_{0.19}As$ layer. A top layer, typically 10 nm of $In_{0.81}Ga_{0.19}As$, is grown on InAs strained quantum well as shown in Figure. 9.1(b). After the quantum well is grown, the substrate is cooled to promote the growth of epitaxial Al (111) [16]. Figure 9.1(c) shows a high-resolution transmission electron microscope (TEM) image of this interface between Al and $In_{0.81}Ga_{0.19}As$, with atomic planes of both crystals clearly visible. All samples reported in this paper have had Al thin films grown and thin films of Al were selectively removed, using Al etchant Transene solution type D, for transport studies of the InAs quantum well. The surface roughness of samples does not change with or without Al thin films confirmed by atomic force microscope images.

Engineering the coupling of the 2DEGs with surface is an important parameter in tuning the proximity effect. The goal is to keep the finite charge distribution at the surface in addition to highest electron mobility and quality of magnetotransport. In pure InAs samples, electron is accumulated at surface due to pinning of the surface Fermi level above the conduction band minimum. The position of the pinning level depends on the crystal direction and the surface treatments [33]. Experiments on $In_xGa_{1-x}As$ predict Schottky barrier height becomes negative, exhibiting an ohmic behavior for $x > 0.8536$. Low temperature transport properties of surface accumulation InAs structures are dominated by surface scattering (low mobility, limited to a 1000–3000 cm^2/Vs) and magnetotransport shows very weak signatures of Shubnikov–de Haas oscillations. Adding a top layer to pure InAs quantum well resolves these issues. The corresponding charge distributions are shown in Figure 9.2(a)–9.2(c). As evidenced by charge distribution calculations in Figure 9.2(a), there is a strong overlap with the surface for the 2 nm InGaAs top layer and the overlap decreases by increasing the top layer thickness where it vanishes near 20 nm of InGaAs top layer shown in Figure 9.2(c). We have studied several samples with various InGaAs top layer thicknesses and have found that a 10 nm top layer thickness yields the highest quality 2DEGs as well as good wave function overlap at the surface as shown in Table 9.1. We have also tested higher barrier top layers such as InAlAs which should be narrower for the same overlap near the surface. The coupling of the 2DEG with the surface can be precisely tuned using the composition and the thickness of top layers.

While the coupling of the 2DEG to surface is an important factor, the 2DEG itself should have a high electron mobility. To understand various sources of scattering, we study the density dependence of the electron mobility. Figure 9.2(d) shows calculated mobility due to scattering from rough interface, alloy scattering and background doping which are all relevant for deep structures (100–200 nm of InAlAs top layer) [31, 32]. The alloy scattering is highly sensitive to the ternary composition, $\mu = \frac{1}{x(1-x)}$; calculation is done following [34]. The background doping N_B is deduced based on the observation

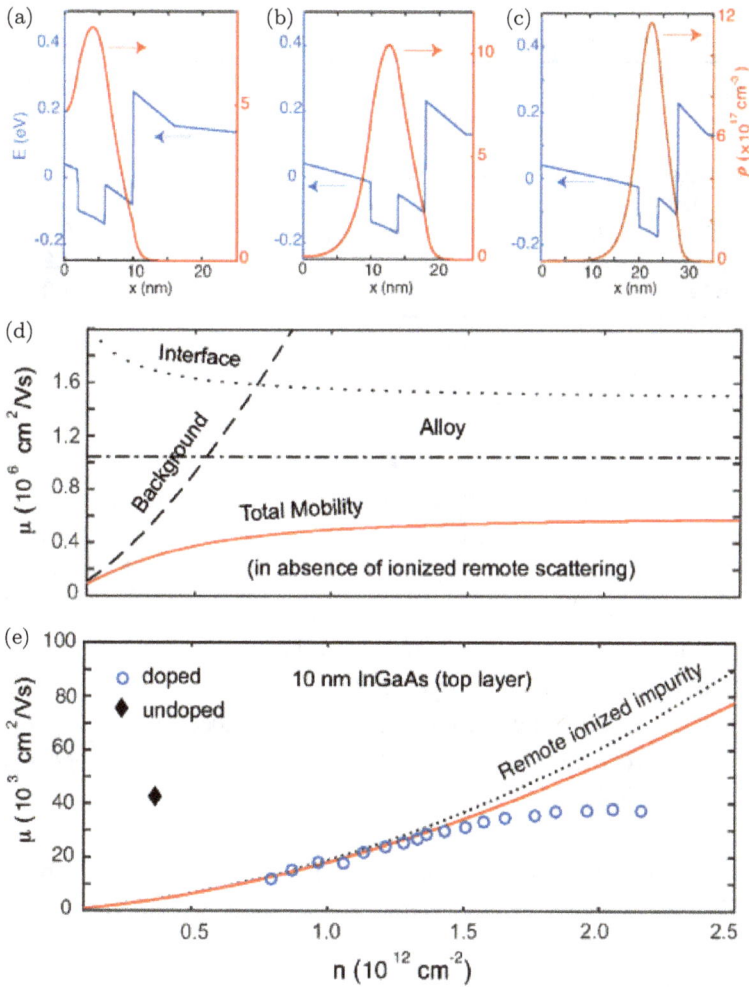

Figure 9.2.   Self-consistent Poisson charge distribution calculations for top layers of (a) 2 nm InGaAs, (b) 10 nm InGaAs and (c) 20 nm. (d) Calculated mobilities from different scattering mechanisms as a function of doping electron density for buried quantum wells. (e) Total theoretical mobility vs. measured mobility for surface quantum wells. Solid red lines correspond to the total calculated mobilities whereas the dash lines correspond to calculated mobilities due to individual scattering mechanisms.

Table 9.1.   Summary of InAs surface quantum wells and their transport properties.

Barrier	d(nm)	N_d(cm^{-2})	$n(10^{11}$cm$^{-2})$	μ(cm^2/Vs)	Bon(T)
InAlAs	5	10^{12}	10	14,000	1.5
InAlAs	2	10^{12}	6.76	8,500	2
InGaAs	2	10^{12}	5.8	6,500	3
InGaAs	10	10^{12}	7.15	14,400	2.25
InGaAs	10	—	3.6	44,000	1.2
InGaAs	20	10^{12}	10	12,570	1.3

Top layer material with indium composition of 0.81, barrier thickness in nm: d, Si doping (cm^{-2}): N_d, Electron density (cm^{-2}): n, Mobility: μ, onset of Shubnikov-de Haas (Tesla): Bon.

that undoped deep structures conduct with density range in few 10^{11}cm^{-2}. Based on series of comparison between calculation and measurements, we estimate $N_B \sim 10^{16}$cm^{-3}. The interface roughness with fluctuation height 0.8 nm adds a cutoff at 1.5×10^6 cm^2/Vs [35]. Using these values, the total mobility is calculated based on the Matthiessen's rule as shown in red in Figure 9.2(d). For electrons confined deeper in the structures (and not near surface), mobility is expected to increase with electron density and saturates around $600,000$ cm^2/Vs similar to earlier studies [31]. However, recent results in similar structures show higher mobilities can be achieved [24] on 120 nm deep structures which may suggest a discrepancy between experiments and alloy scattering model used from [34]. In binary InAs quantum wells the electron mobility can even reach up to $\mu = 2.4 \times 10^6$ cm^2/Vs and signatures of fractional quantum Hall states have been observed [25, 26]. Note that the mobility in our structures is traded to achieve epitaxial superconducting contacts which cannot readily happen in deeper 2DEGs.

For surface quantum wells, the situation is different where surface scattering and remote ionized doping play major roles. The dependence of the mobility on density of our sample is shown in Figure 9.2(e) as open blue circles. The electron mobility varies between 10,000 cm^2/Vs and 40,000 cm^2/Vs. The mobility starts to saturate near 1.6×10^{12} cm^{-2} close to second sub-band occupation (confirmed by calculation). The data can be fitted by $\mu \sim n^{\alpha}$, with $\alpha = 0.9$ in density range below $1.6 \times 10^{12}\,\mathrm{cm}^{-2}$. The contribution of background doping is low in surface quantum wells as seen in Figure 9.2(d). The main scattering mechanism is surface and depends on details of scatter centers. For simplicity, we assume that the surface scattering contributions are similar to remote ionized impurities [36]. We first fit the $44{,}000\,\mathrm{cm}^2/\mathrm{Vs}$ mobility data point from our undoped sample which results in $10^{11}\,\mathrm{cm}^{-2}$ ionized impurity. We use this number and add intentional Si doping, $N_d \sim 7.5 \times 10^{11}\mathrm{cm}^{-2}$ and plot the combined results in Figure 9.2(e). Solid red line corresponds to the total calculated mobility whereas the dash line corresponds to calculated mobility due to remote ionized impurities (both surface and Si doping). We have achieved an enhanced mobility of $44{,}000\,\mathrm{cm}^{-2}/\mathrm{Vs}$ at $n = 3.6 \times 10^{11}\,\mathrm{cm}^{-2}$ without Si doping (with onset of oscillation at 1.2 T). The self-consistent calculation estimates a carrier density of $2.16 \times 10^{11}\,\mathrm{cm}^{-2}$ from background doping and $1 \times 10^{11}\,\mathrm{cm}^{-2}$ from surface with a total density of $3.16 \times 10^{11}\,\mathrm{cm}^{-2}$ which is close to the measured density. This suggests that in optimized structures, the mobility can be increased without intentional doping.

The SOCs in near surface quantum wells are highly tunable since they are confined on one side by an infinite barrier (electron cannot jump out of the surface) and on the other side by a finite barrier of InAlAs. The electric field across the quantum well is forcing the 2DEG to have Rashba SOC. By tuning the gate voltage and changing the slope of the bands one can show Rashba parameter derived from SOC-induced weak antilocalization correction to conductivity varies from zero to 0.1 eVÅ [37].

9.4. Josephson-Field Effect Transistor on Epitaxial Al–InAs Heterostructures

Superconducting devices based on Al–AlO$_x$–Al tunnel junctions are the building block of most Josephson circuits from sensing and high-speed digital electronic devices to transmon qubits with long coherence times. Aluminium-based tunnel junctions have proven to be more reliable and stable than any other material, thanks to the formation of pristine Al–AlO$_x$ interfaces through the natural oxidization of Al. Similarly, Josephson-field effect transistor (JFET) consists of a semiconductor-coupled weak link between two superconducting leads combining the physics of superconductivity with the electrostatic tunability of semiconductors. JFET architecture allows the control of the charge density in the semiconductor and hence the superconducting properties of the weak links. The Andreev reflection at the interface between the superconductor and semiconductor allows an incoming electron is retroreflected as a hole with creation or destruction of a Cooper pair in the superconductor. While the JFETs as a concept have been known for decades, their applicability were severely limited by reduced Cooper pair injection at the superconductor–semiconductor interface. JFET devices were identified as low-power transistors in 1980s [10] but it was realized without a natural and coherent interface between a superconductor and a semiconductor and the device yield and reliability are low. As we described, one can develop a wafer-scale method for the epitaxial growth of thin films of Al on III–V compound heterostructures [37] with atomically flat interfaces. These epitaxial contacts can be made only to electrons confined near to surface where mobility is dominated by surface scattering. Depending on the application, in some cases high electron mobility is desired [19, 38] and in other cases control over the induced gap is called for [20]. As we showed, a 10 nm thick top layer of In$_{0.81}$Ga$_{0.19}$As can achieve both [16, 37].

Josephson junctions are formed by selective removal and gate-stack definition. Al films were selectively etched by Transene type-D solution while for optical studies Al was not grown from the beginning. A 50 nm thick aluminum oxide gate dielectric is then deposited on top of the Josephson junction via atomic layer deposition. Gate electrodes were realized by subsequent deposition of 5 nm of titanium and 70 nm of gold. All measurements were performed inside a cryogen-free refrigerator with base temperature of 20 mK.

Figure 9.3(a) shows a scanning electron microscope image of a finished JFET device. An Al stripe of 100 nm by 4 μm is etched where InAs is exposed and junction is formed. The dielectric is deposited

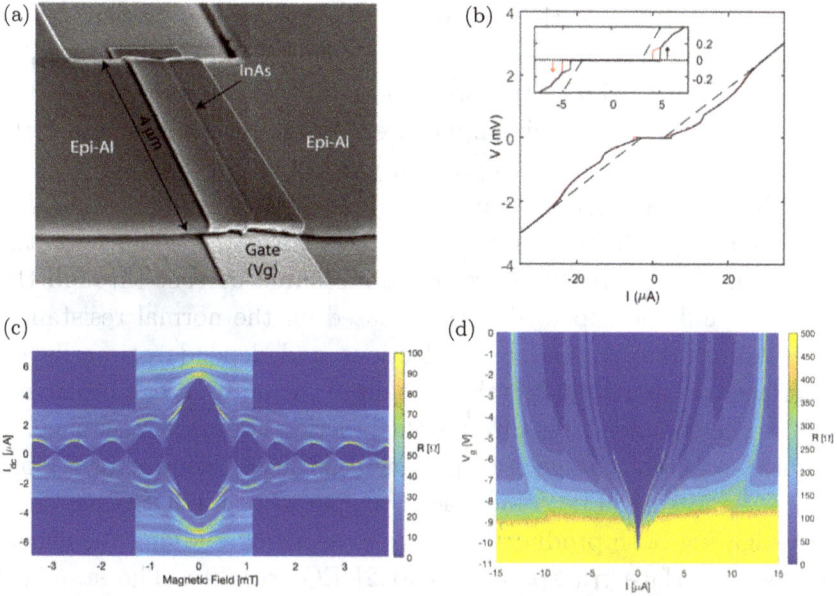

Figure 9.3. (a) Scanning electron microscope image of a JFET. (b) Voltage–current curve for 100 nm. Inset shows the critical current when the junction switches from zero-voltage state to normal. Dashed line shows the extrapolation of normal resistance to determine the excess current. (c) Fraunhofer pattern of a Josephson junction shown in part (a). (d) Differential resistance of a junction with width of 100 nm. Dark blue represents zero resistance. All measurements performed at 20 mK in a dilution fridge.

over the whole chip and a finger gate is fabricated over the InAs stripe as shown. When considering the properties of a JFET, both transparency of the InAs–Al interface and scattering within the 2DEG need to be considered. Scattering in the 2DEG determines whether transport through the junction is diffusive or ballistic, generally characterized by the mean free path l_e in comparison with junction length, L. Magnetotransport reveals typical mean free paths $l_e \sim 200$ nm for our samples [39]. Our Al layer thickness is close to 12 nm and hence the critical temperature is enhanced compared to bulk value (1.2 K) and is found to be $T_c = 1.48\,K$. Using the relation $\Delta_{Al} = 1.75 k_B T_c$ we find $\Delta_{Al} = 223\,\mu V$. Al thickness variation results in 10% variation in magnetic field measured across the wafer. The critical field of these films is near 100 mT. From

Δ_{Al} we can estimate the superconducting coherence length in our samples given by $\xi_0 = \hbar\nu_F/(\pi\Delta)$ which yields $\xi_0 = 774\,\text{nm}$ for density of $7 \times 10^{11}\,\text{cm}^{-2}$. From this we expect all devices to approach the dirty limit $(\xi_0 \gg l_e)$. This implies we should also consider the dirty coherence length $\xi_{0,d} = \sqrt{\xi_0 l_e}$ which yields $400\,\text{nm}$.

These hybrid systems can be characterized by study of the $I_c R_N$ and $I_{ex} R_N$ products, where R_N is the normal resistance of the JJ. I_{ex} is the difference between the measured current through the junction and the expected current based on the normal resistance. This occurs due to Andreev reflections and depends primarily on interface transparency. I_c is the amount of current that can be carried by Andreev bound states through the junction with zero resistance. I_c requires coherent charge transport across the semiconductor region, so it is a measure of both interface transparency and 2DEG mobility. Investigating both products provides a means of studying the effects of both interface transparency and 2DEG mobility. The standard figure of merit for a JFET is $I_c R_N$ to remove the geometrical aspect of the device. As the width increases there will be a higher number of channels contributing where one expect a higher current but also a lower resistance. It should be noted that experimentally there is a distinction between I_C and the measured switching current of a junction. In an ideal system, these quantities would be identical, but it is commonly seen that finite temperature or noise can cause the junction to prematurely switch. This leads to measuring an observed switching current lower than I_C. For simplicity in the analysis, we assume that they are equal in our devices while knowing this can lead to an underestimation of our I_C. The critical current can be related to the gap by the formula $I_C R_N = \alpha\Delta_0/e$. From Figure 9.3(b) inset, we observe $I_C R_N = 486\,\mu V$. This value can be compared to theoretical values for fully transparent junctions in the short diffusive and ballistic limits, for which α are $1.32(\pi/2)$ and π, respectively [40, 41] when JJs are deep in ballistic $(l_e \gg L)$ or diffusive regime$(L \gg l_e)$. This suggest that our JFET is in a quasi-ballistic regime with $I_C R_N$ being 69% of the ballistic limit and 105% of the diffusive limit. It is clear that $100\,\text{nm}$ JFET is close to the short ballistic regime.

High interface transparency corresponds to a high probability of Andreev reflection at the interface. Since the semiconductor extends under the superconductor regions, the interface should be highly transparent due to the large area of contact and *in-situ* epitaxial Al growth [42]. The Andreev process that carries the supercurrent across the semiconductor region is characterized by the excess current (I_{ex}) through the junction $I_{ex} = I - V/R_N$ [43]. Excess current does not require coherent charge transport across the junction as it follows simply from charge conservation at one interface. This allows for the excess current to be calculated by extrapolating from the high current normal regime to zero voltage as shown in Figure 9.3(b) with dashed lines. The excess current in our JFET is found to be $I_{ex} = 3.5\,\mu A$ for 100 nm JFET. As we mentioned when considering interface quality, the more relevant quantity is the product $I_{ex}R_N$. The product $I_{ex}R_N$ can be compared to the superconducting gap with the relation $I_{ex}R_N = \alpha'\Delta_0/e$. In the case of a fully transparent S–Sm interface $\alpha' = 1.467$ for a diffusive junction [44] and $\alpha'=8/3$ for a ballistic junction [43]. For our sample, $I_{ex}R_N = 340\,\mu V$. Comparing this to the ballistic and diffusive limits, we see that our values are 57% of ballistic limit and at 104% actually slightly exceeds the diffusive value for a 100 nm JFET.

Figure 9.3(c) shows the resistance as a function of perpendicular magnetic field on our JFET. The critical current (zero-resistance) shows a sinc function which qualitatively matches the expected model where integer numbers of flux quanta penetrate the normal region. The deviation in node spacing is attributed to flux focusing by 20%. An inverse Fourier transform of the pattern shows a relatively uniform current distribution across the junction. The JFET is also equipped with a metallic gate marked as V_g that allows the junction resistance and critical current o be varied. Figure 9.3(d) shows the differential resistance in color as current bias and gate voltage are changed. The critical current can be tuned from $5\,\mu A$ to zero using by sweeping the V_g from zero to -11 V. The high voltage needed here is due to thick dielectric thickness. This field effect control of the superconducting current allows for a wide range of applications from logic to fast digital electronics such as single flux quantum (SFQ) circuits.

We generally find that 2DEG mobility and the inferred interface transparency seem to be closely related, possibly due to the fact that the mobility of surface 2DEGs has been found to be dominated by surface scattering [37]. This implies that the same impurities affecting the 2DEG mobility will also dominate degradation of the interface in the case of *in-situ* epitaxial growth, which would otherwise produce a transparent interface. In the high mobility samples, we find product $I_C R_N / \Delta_0 \sim 2.2$. Remarkably, we have shown that $I_{ex} R_N / \Delta_0 \sim 1.5$ to be independent of both junction length and gate voltage. These junctions can potentially replace the Al-based tunnel junctions and be embedded in qubit architectures, as recently shown in [8] on InAs nanowires and planar Josephson junctions. The epitaxial superconductor–semiconductor material provide a new platform for not only studying mesoscopic superconductivity but also topological phases of matter. Recent reports show proximitized electrons in a Josephson junction with Rashba spin–orbit coupling and showed that they can enter a topological superconducting phase in the presence of a Zeeman field, similar to the proximitized nanowires, where the Majorana bound state appears at the ends of the junction [45]. The major advantage of this two-dimensional junctions is that the phase transition can be tuned using the gate voltage and superconducting phase [46]. This allows for more complicated networks that could support fusion, braiding, and larger-scale Majorana networks.

Bibliography

[1] V. Mourik, K. Zuo, S. Frolov, S. R. Plissard, E. P. Bakkers, and L. P. Kouwenhoven, *Science*, **336**, 1003 (2012).
[2] P. Krogstrup *et al.*, *Nat. Mater.*, **14**, 400 (2015).
[3] J. W. Orton, and T. Foxon, *Molecular Beam Epitaxy: A Short History* (Oxford University Press, 2015).
[4] J. Alicea, *Rep. Prog. Phys.*, **75**, 076501 (2012).
[5] S. R. Elliott, and M. Franz, *Rev. Mod. Phys.*, **87**, 137 (2015).
[6] Z. Qi *et al.*, *Phys. Rev. B*, **97**, 134518 (2018).
[7] L. Casparis *et al.*, *Nat. Nanotechnol.*, **13**, 915 (2018).

[8] T. W. Larsen, K. D. Petersson, F. Kuemmeth, T. S. Jespersen, P. Krogstrup, J. Nygård, and C. M. Marcus, *Phys. Rev. Lett.*, **115**, 127001 (2015).

[9] Z. Wan, *et al.*, *Nat. Comm.*, **6**, 7426 (2015).

[10] T. D. Clark, R. J. Prance, and A. D. C. Grassie, *J. Appl. Phys.*, **51**, 2736 (1980).

[11] A. Richter, M. Koch, T. Matsuyama, and U. Merkt, *Supercond. Sci. Technol.*, **12**, 874 (1999).

[12] H. Kroemer, C. Nguyen, and E. L. Hu, *Solid-State Electron.*, **37**, 1021 (1994).

[13] H. Takayanagi, and T. Akazaki, *Solid State Commun.*, **96**, 815 (1995).

[14] Y.-J. Doh *et al.*, *Science*, **309**, 272 (2005).

[15] W. Chang *et al.*, *Nature Nanotechnol.*, **10**, 232 (2015).

[16] J. Shabani *et al.*, *Phys. Rev. B*, **93**, 155402 (2016).

[17] I. Zutic, J. Fabian, and S. Das Sarma, *Rev. Mod. Phys.*, **76**, 323 (2004).

[18] R. M. Lutchyn, E. P. A. M. Bakkers, L. P. Kouwenhoven, P. Krogstrup, C. M. Marcus, and Y. Oreg, *Nature Rev. Mat.*, **3**, 52 (2018).

[19] J. D. Sau, S. Tewari, and S. Das Sarma, *Phys. Rev. B*, **85**, 064512 (2012).

[20] W. S. Cole, S. Das Sarma, and T. D. Stanescu, *Phys. Rev. B*, **82**, 174511 (2015).

[21] J. Nitta, T. Akazaki, H. Takayanagi, and K. Arai, *Phys. Rev. B*, **46**, 14286 (1992).

[22] L. C. Mur, C. J. P. M. Harmans, J. E. Mooij, J. F. Carlin, A. Rudra, and M. Ilegems, *Phys. Rev. B*, **54**, R2327 (1996).

[23] J. P. Heida, B. J. van Wees, T. M. Klapwijk, and G. Borghs, *Phys. Rev. B*, **57**, R5618 (1998).

[24] A. T. Hatke, T. Wang, C. Thomas, G. Gardner, and M. Manfra, *Appl. Phys. Lett.*, **111**, 142106 (2017).

[25] T. Tschirky, S. Mueller, C. A. Lehner, S. Fält, T. Ihn, K. Ensslin, and W. Wegscheider, *Phys. Rev. B*, **95**, 115304 (2017).

[26] M. K. Ma, M. S. Hossain, K. A. Villegas Rosales, H. Deng, T. Tschirky, W. Wegscheider, and M. Shayegan, *Phys. Rev. B*, **96**, 241301 (2017).

[27] C. Lehnert, Konrad W. *Nonequilibrium dynamics in mesoscopic superconductor-semiconductor-superconductor junctions.* University of California, Santa Barbara. 1999. Thesis.

[28] H. J. Suominen, M. Kjaergaard, A. R. Hamilton, J. Shabani, C. J. Palmstrøm, C. M. Marcus, and F. Nichele, *Phys. Rev. Lett.*, **119**, 176805 (2017).

[29] F. Nichele *et al.*, *Phys. Rev. Lett.*, **119**, 136803 (2017).

[30] X.Wallart, J. Lastennet, D. Vignaud, and F. Mollot, *Appl. Phys. Lett.*, **87**, 043504 (2005).

[31] J. Shabani, A. P. McFadden, B. Shojaei, and C. J. Palmstørm, *Appl. Phys. Lett.*, **105**, 262105 (2014).

[32] J. Shabani, S. Das Sarma, and C. J. Palmstørm, *Phys. Rev. B*, **90**, 161303 (2014).

[33] L. Canali, J. Wildoer, O. Kerkhof, and L. Kouwenhoven, *Appl Phys A*, **66**, 113 (1998).

[34] D. Chattopadhyay, *Phys. Rev. B*, **31**, 1145 (1985).

[35] B. Shojaei, P. J. J. O'Malley, J. Shabani, P. Roushan, B. D. Schultz, R. M. Lutchyn, C. Nayak, J. M. Martinis, and C. J. Palmstørm, *Phys. Rev. B*, **93**, 075302 (2016).

[36] A. Gold, *Phys. Rev. B*, **35**, 723 (1987).

[37] K. S. Wickramasinghe, W. Mayer, J. Yuan, T. Nyugen, L. Jiao, V. Manucharyan, and J. Shabani, *Appl. Phys. Lett.*, **113**, 262104 (2018).

[38] A. Haim, and A. Stern, *Phys. Rev. Lett.*, **122**, 126801 (2019).

[39] W. Mayer *et al.*, *Appl. Phys. Lett.*, **114**, 103104 (2019).

[40] V. Ambegaokar, and A. Baratoff, *Phys. Rev. Lett.*, **10**, 486 (1963).

[41] K. K. Likharev, *Rev. Mod. Phys.*, **51**, 101 (1979).

[42] M. Kjaergaard, H. J. Suominen, M. P. Nowak, A. R. Akhmerov, J. Shabani, C. J. Palmstrøm, F. Nichele, and C. M. Marcus, *Phys. Rev. Appl.*, **7**, 034029 (2017).

[43] G. E. Blonder, M. Tinkham, and T. M. Klapwijk, *Phys. Rev. B*, **25**, 4515 (1982).

[44] I. O. Kulik, and A. N. Omel'yanchuk, *JETP Lett.*, **21**, 96 (1975).

[45] W. Mayer *et al.*, (2019), arXiv:1906.01179.

[46] T. Zhou *et al.*, (2019), *Phys. Rev. Lett.*, **124**, 137001 (2020).

Chapter 10

High-Pressure Synthesis Approaches to Quantum Materials

Valentin Taufour

Department of Physics and Astronomy,
University of California Davis, Davis CA 95616
vtaufour@ucdavis.edu

10.1. Introduction

For the synthesis and discovery of new materials, the use of high-pressure opens a new phase space which remains relatively unexplored to date. There are several reviews on high-pressure synthesis and new material discoveries at high pressures [1–12]. In this chapter, we describe three selected advantages of high-pressure effects that can be harnessed for the growth of new quantum materials. Namely, we illustrate the role of high pressure in (i) stabilizing new crystal structures, (ii) increasing the substitution level, and (iii) stabilizing liquid solutions for new synthesis approaches. We illustrate each cases with recent examples related to quantum materials such as superconductors. We also describe a few technical aspects of high-pressure synthesis, focused on the cubic–anvil technique, to illustrate with a specific example the kinds of skills and tricks that are necessary to perform experiments under high-pressure, high-temperature conditions.

10.2. High-pressure Influence on Crystal Structures: Stabilizing New Crystal Structures

One of the main interest of high-pressure is that it can stabilize crystal structures otherwise unstable at ambient pressure. After

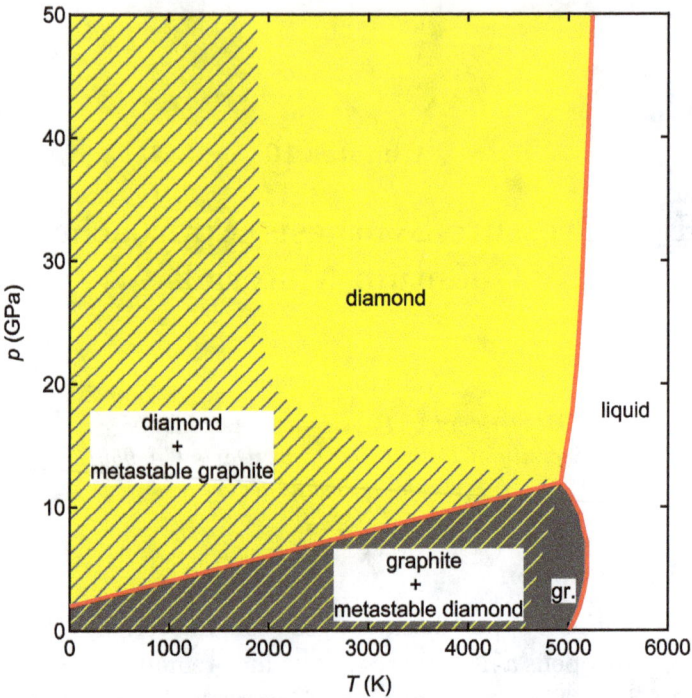

Figure 10.1.   The pressure–temperature phase diagram of carbon proposed by Bundy in 1996 [13].

being synthesized at high pressure, a material can remain stable at ambient pressure due to large enough transformation barriers. A famous example is diamond. The pressure–temperature phase diagram of carbon, as proposed by Bundy in 1996 is illustrated in Figure 10.1. At ambient pressure, the stable structure of solid carbon is not diamond, but graphite. Natural diamonds form in Earth's upper mantle at very high temperatures and pressures, and reach the surface during volcanic eruptions. The large energy barriers between the diamond structure and the graphite structure prevent diamond to convert back into graphite. Diamond is said to be metastable, or kinetically stable, but not thermodynamically stable.

High-pressure synthesis can be used to synthesize materials such as diamond whose stability is increased under high-pressure conditions. However, the search for new materials stable under high

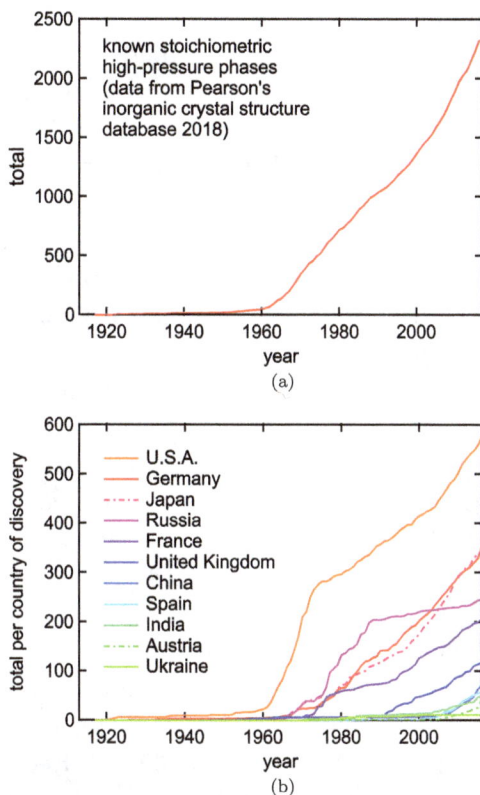

Figure 10.2. (a) Time evolution of the total number of known high-pressure phases (known crystal structure) of stoichiometric inorganic compounds. The first structure determination of a high-pressure phase was the carbon diamond structure by the Braggs in 1913 [20]. (b) Time evolution for the main countries of discovery.

pressure can quickly become a random fishing expedition unless it is supported by theoretical predictions [14–19]. Figure 10.2(a) shows the historic evolution of high-pressure phases discovery. As for most scientific fields, the discovery rate strongly accelerated in the 1960's. For inorganic compounds, more than 2300 stoichiometric high-pressure phases have been reported so far. Since 2005, an average of 66 new high-pressure phases are discovered every year (see Figure 10.2(a)), mostly in the USA, Germany, Japan, and a rapid increase in China (see Figure 10.2(b)).

A recent example in the field of superconductivity is the discovery of FeB_4. Various theoretical studies predicted the possible synthesis of FeB_4 [21], and others tetraborides such as ReB_4 [22, 23], TaB_4 [24, 25], and IrB_4 [26] under high-pressure conditions. The new compound FeB_4 was predicted to crystallize in the orthorhombic structure [space group $Pnnm$ (58)] with outstanding hardness, as well as superconductivity [21]. High-pressure synthesis was suggested as a possible route to obtain FeB_4 because of the predicted stabilization of the structure under pressure [27]. In 2013, FeB_4 was successfully synthesized under pressures above $10\,GPa$ and temperatures above $1500\,K$ [28], and the predicted orthorhombic structure as well as the superconductivity ($T_{sc} = 2.9\,K$) were confirmed.

Substantial progress has also been made in the high-pressure synthesis of hydrides [29–32]. Pressure can induce the formation of many new hydride compounds, with potential applications as reducing agents, precursors, energy storage, and sensors [31]. Hydrides are also relevant to the study of magnetism [30] and superconductivity [33–35]. Hydrogen atoms in hydrides have a high vibrational frequency which is favorable for phonon-mediated superconductivity. This property has motivated the search for hydrogen rich alloys which could be stabilized under high-pressure [35–47]. Recently, Alexander P. Drozdov *et al.* found that H_3S is a superconductor with a record-high T_{sc} of up to $203\,K$ under a pressure of $155\,GPa$ [48]. They broke that record again when reporting a T_{sc} of $250\,K$ at $170\,GPa$ for LaH_{10} [49]. The high superconducting transition temperature in LaH_{10} was also independently discovered by Maddury Somayazulu *et al.* [50]. Isotope effect measurements confirm that the superconductivity originates from electron–phonon coupling [49]. In both cases, the superconducting sample is synthesized under high pressure and decomposes upon pressure release.

10.3. High-Pressure Influence on Solid Solubility: Increasing the Substitution Level

Another interest of high-pressure is that it can modify the solid solubility of a particular element in a crystal. A relatively

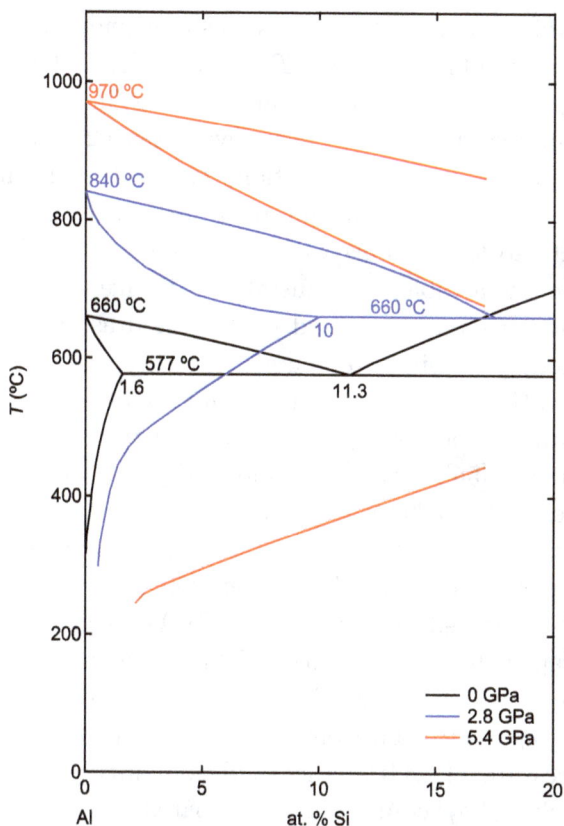

Figure 10.3. The pressure evolution of the Al-rich region of the Al–Si binary phase diagram reported by Mii *et al.* in 1976 [51].

well-documented example is shown in Figure 10.3 with the aluminum-rich region of the aluminum–silicon (Al–Si) binary phase diagram and its evolution under high pressure. At ambient pressure, the solubility limit of Si in Al is 1.6 at.% at the eutectic temperature. Under a pressure of 2.8 GPa, the maximum solid solubility increases to 10 at.%, and exceeds 20 at.% for pressures above 5.4 GPa [51].

Substitutions of an element by another is often used to modify the properties of materials. Practically, an increased solid solubility limit offers a larger range to tune the physical properties of a material. Staying with our example of the Al–Si system, it turns out that increasing the amount of Si substitutions in Al enables an increase of the superconducting transition temperature. Pure Al becomes superconducting below $T_{sc} = 1.175$ K, and the transition temperature increases with Si substitutions in $Al_{1-x}Si_x$ alloys. Using high-pressure synthesis, alloys with $x \approx 0.2$ can be prepared and their superconducting transition temperature is increased to $T_{sc} \approx 11$ K [52, 53]. At ambient pressure, the lattice is unstable for $x > 0.016$, so the crystals with $x \approx 0.2$ synthesized under high pressure are metastable. The instability is likely to enhance the electron–phonon interactions and induce the drastic increase of the superconducting transition temperature [54].

Similarly, the increase of solid solubility under high pressure was used to synthesize better superconductors in the family of Fe-based superconductors [55]. The compound $SrFe_2As_2$ becomes antiferromagnetic below $T_N = 200$ K [56]. Superconductivity can be observed when the magnetism is weakened by applying pressure [57–59] or using various chemical substitutions [60–67]. Superconductivity can even be induced by exposing $SrFe_2As_2$ epitaxial thin films to water vapor [68]. In the case of La substitutions for Sr, the solid solubility of La in $SrFe_2As_2$ is small, and synthesis attempts result in segregation of La from the $SrFe_2As_2$ phase [55]. However, under a pressure of 2 GPa, the solid solubility is increased and crystals of $Sr_{1-x}La_xFe_2As_2$ can be prepared with up to $x = 0.6$. Superconductivity is observed for $x > 0.3$ and the superconducting transition temperature reaches $T_{sc} = 22$ K for $x = 0.4$ with a magnetic shielding volume fraction of 70% [55].

The same advantage of high-pressure synthesis was also used on the compound $CaFe_2As_2$. Traces of superconductivity can be observed when substituting Ca by rare-earth elements [69–71]. Figure 10.4 shows the substitution level of Pr determined from wavelength-dispersive X-ray spectroscopy (WDS) as a function of the nominal concentration used in the synthesis. At ambient pressure, the solid solubility is limited, and the amount of substitution saturates [69]. In a collaboration between Ames Laboratory and the University of Maryland, the present author and his collaborators used high-pressure synthesis to prepare samples of $Ca_{1-x}Pr_xFe_2As_2$ with larger substitution levels. Using a pressure of 3.3 GPa and

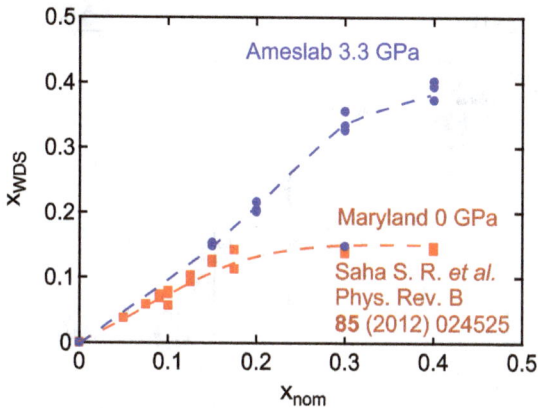

Figure 10.4. Chemical analysis of Pr concentration x in $Ca_{1-x}Pr_xFe_2As_2$ obtained from wavelength-dispersive X-ray spectroscopy (WDS) vs. the nominal concentration used in the synthesis.

temperature of $800°C$, the substitution level can be more than doubled compared to the value at ambient pressure (see Figure 10.4).

It is important to note that pressure does not always increase the solid solubility. In the examples given here, the substitution induces a reduction of the lattice parameters, and the ionic size mismatch limits the solid solubility. By applying pressure during the synthesis, the lattice is compressed and the mismatch is reduced, allowing for an increase of the solid solubility.

10.4. High-Pressure Influence on Dew and Bubble Curves: Stabilizing Liquid Solutions for New Synthesis Approaches

Another important interest of high pressure is that it can increase the temperature of the dew and bubble curves, enabling new synthesis approaches. The states with higher density are stabilized by the application of pressure, which will generally increase the melting and boiling temperatures. In a multicomponent system, the vapor and the liquid can coexist while having different composition. The dew curve gives the temperature at which the first drop of liquid appears upon cooling a gas at a given composition, and the bubble curve gives the

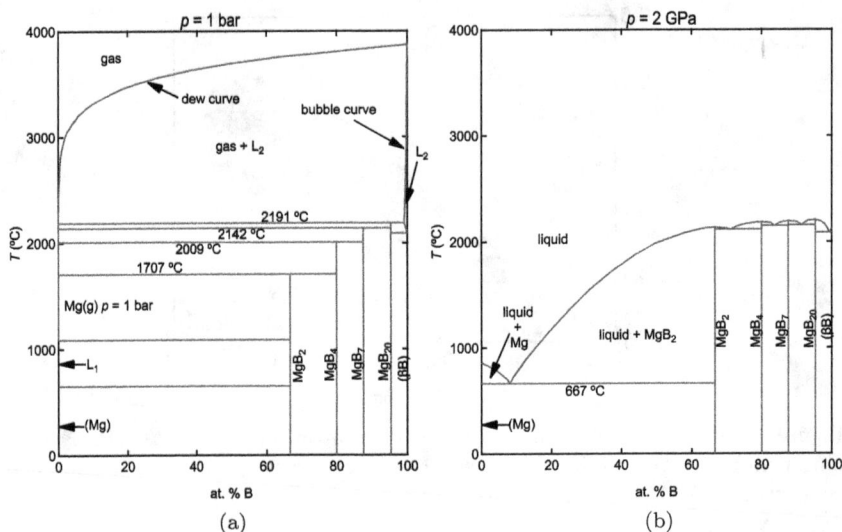

Figure 10.5. Calculated compositional phase diagram of the Mg–B binary system (a) at ambient pressure [72, 73] and (b) at 2 GPa [74].

temperature at which the first bubble of vapor forms when heating a liquid at a given composition.

In a binary system such as magnesium–boron (Mg–B), the boiling point of Mg (1090°C [75]) is much lower than the boiling point ($\sim$4000°C [75]) and even the melting point (2075°C [75]) of B and of other compounds such as MgB_2, MgB_4, etc. The bubble curve and the liquid region are therefore "hidden" as the solid compounds will coexist with a Mg rich gas (see the diagram in Figure 10.5(a) calculated at ambient pressure [72, 73]). Under pressure, the temperature of the dew and bubble curves will increase much more than the melting temperatures of the compounds. This can open a region of liquid phase in the diagram. For example, the compositional phase diagram of the Mg–B system calculated at 2 GPa in [74] is reproduced in Figure 10.5(b).

The opening of the liquid region under pressure allows for a new approach to grow single crystals of MgB_2: the solution growth technique with an excess Mg is possible under high pressure [76–82]. MgB_2 has attracted a lot of attention in the early 2000s after the discovery of multiband superconductivity below $T_{sc} = 49$ K [83].

Recent calculations predicted the existence of topological Dirac nodal lines as well as topological surface states connecting them [84]. These properties make MgB_2 a promising compound to study topological superconductivity and revive the need for high-quality single crystals.

In practice, elemental Mg and B are placed in a crucible which is taken to high pressures and high temperatures. When boron nitride (BN) crucibles are used, BN reacts with the melt and hexagonal-BN single crystals are grown together with MgB_2 single crystals [85]. In order to control the isotopic purity of boron and avoid contamination from the BN crucibles made from natural abundance B, we used MgO crucibles to grow single crystals of both $Mg^{11}B_2$ and $Mg^{10}B_2$ [81]. Under these conditions, only single crystals of MgB_2 form and the isotopic purity of B can be preserved.

The use of high pressure with volatile elements prevents evaporation and losses, and can open access to a liquidus surface as illustrated in the case of the Mg–B system. Another advantage arises when using the high-pressure anvil technique: it can also increase safety compared with synthesis techniques using sealed ampoules. In the high-pressure technique, the sample is confined in a closed container supported by the anvils, whereas explosions of ampoules may lead to contamination of the laboratory with poisonous materials [9].

10.5. Technical Aspects: Case of a Cubic–Anvil High-Pressure Furnace

To characterize quantum materials, measurements of physical properties such as magnetic susceptibility, electrical conductivity, or specific heat are necessary. Conventional techniques usually require samples with size greater than ~ 1 mm^3, and single crystals are better than polycrystals for a deeper understanding of the physical properties such as anisotropic superconducting gap symmetry, magneto-crystalline anisotropy, or even simple electric transport properties. Among the variety of large-volume high-pressure apparatus, the cubic–anvil apparatus is commonly used for pressures below ~ 5 GPa. The pressure is applied along the six faces of a cubic furnace chamber as illustrated in Figure 10.6. The pressure is generated by applying a

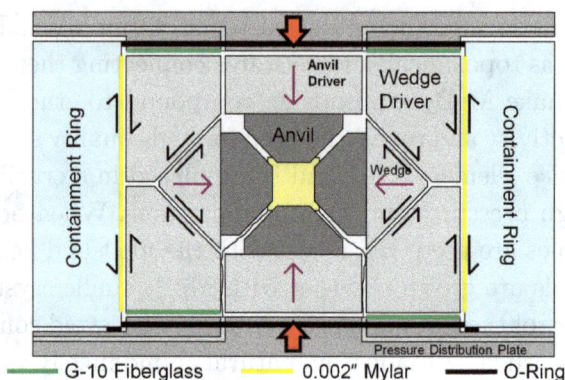

Figure 10.6. Schematic illustration of the cross-section of the pressure chamber [86].

vertical force on a pressure distribution plate, which in turn applies a load onto hardened steel wedge drivers and the top and bottom tungsten carbide anvils. The wedge drivers transmit a load onto the four wedges, which apply forces on the side anvils and the cubic furnace chamber in the horizontal plane. Although the compressive strength of tungsten carbide is in the range 4–6GPa depending on the grade, its transverse rupture strength is in the range 1.5–2.5 GPa and it is classified as a brittle material since it exhibits little or no plastic deformation preceding the initiation of a crack and total failure. This means that a careful experimental setup is necessary to obtain pressures in the range of 3–5 GPa. To avoid a stick-slip motion of the anvils, the wedge drivers are lubricated using solid lubricant such as molybdenum disulfide (Figure 10.7). Mylar sheets with PTFE lubricant are used on the side to reduce friction between drivers and the containment ring. It is also important to keep the anvils level during the assembly to ensure a symmetric pressure distribution straight from the beginning of the pressure loading.

The growth of high-quality samples often requires precise *in-situ* control of the temperature, but measuring the temperature under high-pressure conditions constitute another challenge for high-pressure synthesis. Figure 10.8(a) shows the schematic cross-section of the furnace cube assembly. The thermocouple junction is

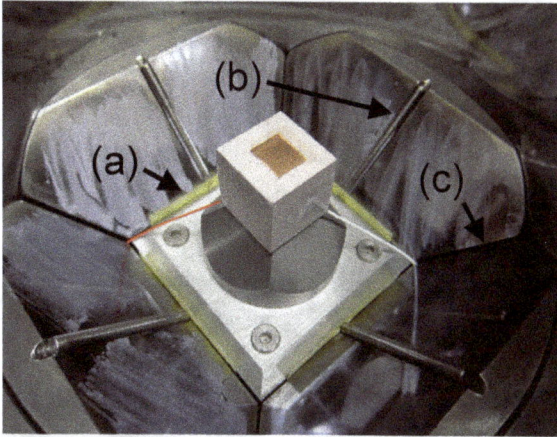

Figure 10.7.   Inside view of the pressure chamber with the bottom anvil and four bottom wedge drivers coated with molybdenum disulfide powder. G10 mini spacers (a), and dowel (b) are assembled. Space between the wedges (c) is made so that the thermocouple wires can freely move [86].

located just below the sample space, through a hole drilled into the pyrophyllite cube. Electrical short circuit between the thermocouple wires and the heater is avoided by using alumina tubing. In addition to making a strong thermocouple junction, special care must be used to prevent the thermocouple wires from breaking during the high-pressure loading. For example, a bend in the thermocouple wire along a face of the cube and secured with Scotch tape can add additional flexibility and reduce the risk of wire rupture (Figure 10.8(b)).

Even when applying all kinds of experimental tricks, high-pressure synthesis remains very challenging and clearly owns its place as a science at extreme conditions. Every detail counts and each improvement contributes to increasing the success rate of the synthesis, and to pushing the pressure and temperature limits always further.

In summary, high-pressure synthesis presents many key advantages for the discovery and study of new quantum materials. In contrast with the complexity of the experimental techniques, the effects of pressure can be modeled well in theoretical calculations, and recent progress in materials simulations provide very promising

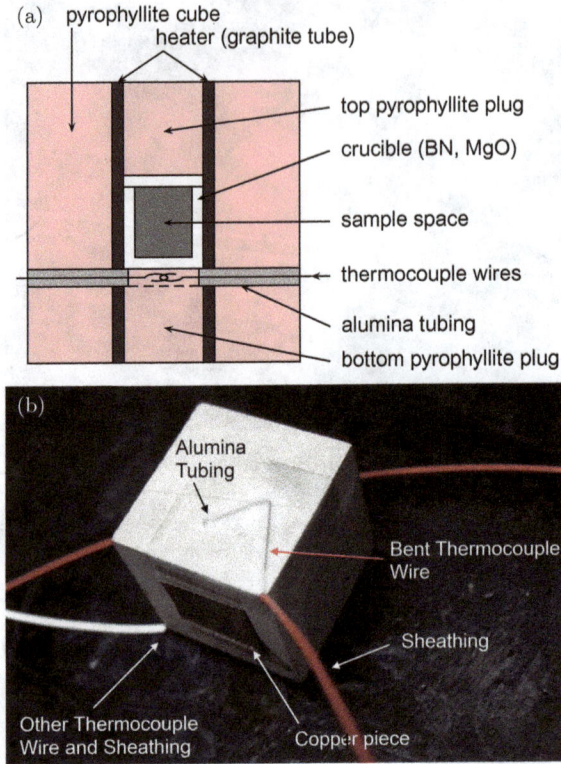

Figure 10.8.   (a) Schematic illustration of the cross-section of the furnace cube assembly. (b) Assembled pyrophyllite cube [86].

tools to guide the synthesis of new quantum materials. High-pressure synthesis is uniquely positioned to become a field in which materials are mostly discovered after theoretical predictions. Because of the enormous phase space opened by the thermodynamic parameter pressure, high-pressure synthesis approaches are expected to develop further and significantly advance the discovery and study of new quantum materials.

Acknowledgments

The author would like to thank Paul Canfield and his group at the Ames Laboratory for teaching the use of a cubic–anvil high-pressure furnace (Rockland Research Corp.). Special thanks to Stella

K. Kim, Soham Manni, and Hyunsoo Kim for their collaboration on the synthesis of MgB_2 single crystals, as well as Tyler Drye and Johnpierre Paglione for their collaboration on the synthesis of $Ca_{1-x}Pr_xFe_2As_2$ crystals.

Bibliography

[1] M. Takano, and A. Onodera, *Curr. Opinion Solid Stat. Mater. Sci.*, **2**, 166 (1997).

[2] J. Badding, *Ann. Rev. Mater. Sci.*, **28**, 631 (1998).

[3] G. Demazeau, *Zeit. Naturforschung Sect. B*, **61**, 799 (2006).

[4] V. V. Brazhkin, *High Pressure Res.*, **27**, 333 (2007).

[5] G. Demazeau, *C. R. Chimie*, **12**, 933 (2009).

[6] P. F. McMillan, in *High-Pressure Crystallography*, E. Boldyreva and P. Dera (eds.) (Springer Netherlands, 2010), pp. 373–383.

[7] C. Q. Jin, Q. Q. Liu, Y. W. Long, J. G. Zhao, Y. X. Lu, L. X. Yang, J. L. Zhu, and X. C. Wang, *Phys. Status Solidi (A)*, **207**, 2750 (2010).

[8] X. Liu, in *Modern Inorganic Synthetic Chemistry*, edited by R. Xu, W. Pang, and Huo, Q. (eds.) (Elsevier Science BV, 2011), pp. 97–128.

[9] J. Karpinski, *Philos. Mag.*, **92**, 2662 (2012).

[10] V. D. Blank, and E. I. Estrin, in *Phase Transitions in Solids Under High Pressure* (CRC Press, 2014), pp. 424–435.

[11] L. Zhang, Y. Wang, J. Lv, and Y. Ma, *Nature Rev. Mat.*, **2**, 17005 (2017).

[12] J. Walsh, and D. E. Freedman, *Acc. Chem. Res.*, **51**, 1315 (2018).

[13] F. Bundy, W. Bassett, M. Weathers, R. Hemley, H. Mao, and A. Goncharov, *Carbon*, **34**, 141 (1996).

[14] A. R. Oganov, and C. W. Glass, *J. Chem. Phys.*, **124**, 244704 (2006).

[15] Y. Wang, J. Lv, L. Zhu, and Y. Ma, *Phys. Rev. B*, **82**, 094116 (2010).

[16] C. J. Pickard, and R. J. Needs, *J. Phys.: Condens. Matter*, **23**, 053201 (2011).

[17] Y. Wang, and Y. Ma, *J. Chem. Phys.*, **140**, 040901 (2014).

[18] Y. Hinuma, T. Hatakeyama, Y. Kumagai, L. A. Burton, H. Sato, Y. Muraba, S. Iimura, H. Hiramatsu, I. Tanaka, H. Hosono, and F. Oba, *Nat. Commun.*, **7**, 11962 (2016).

[19] M. Amsler, V. I. Hegde, S. D. Jacobsen, and C. Wolverton, *Phys. Rev. X*, **8**, 040901 (2018).

[20] H. W. Bragg, and L. W. Bragg, *Proc. Royal Soc. of London. Ser. A*, **89**, 277 (1913).

[21] A. N. Kolmogorov, S. Shah, E. R. Margine, A. F. Bialon, T. Hammerschmidt, and R. Drautz, *Phys. Rev. Lett.*, **105**, 217003 (2010).

[22] B. Wang, D. Y. Wang, and Y. X. Wang, *J. Alloys Compd.*, **573**, 20 (2013).

[23] X. Zhao, M. C. Nguyen, C.-Z. Wang, and K.-M. Ho, *J. Phys.: Condens. Matter*, **26**, 455401 (2014).

[24] B. Chu, D. Li, K. Bao, F. Tian, D. Duan, X. Sha, P. Hou, Y. Liu, H. Zhang, B. Liu, and T. Cui, *J. Alloys Compd.*, **617**, 660 (2014).

[25] S. Wei, D. Li, Y. Lv, Z. Liu, C. Xu, F. Tian, D. Duan, B. Liu, and T. Cui, *Phys. Chem. Chem. Phys.*, **18**, 18074 (2016).

[26] X. Li, H. Wang, J. Lv, and Z. Liu, *Phys. Chem. Chem. Phys.*, **18**, 12569 (2016).

[27] A. F. Bialon, T. Hammerschmidt, R. Drautz, S. Shah, R. Margine, and A. N. Kolmogorov, *Appl. Phys. Lett.*, **98**, 081901 (2011).

[28] H. Gou, N. Dubrovinskaia, E. Bykova, A. A. Tsirlin, D. Kasinathan, W. Schnelle, A. Richter, M. Merlini, M. Hanfland, A. M. Abakumov, D. Batuk, G. Van Tendeloo, Y. Nakajima, A. N. Kolmogorov, and L. Dubrovinsky, *Phys. Rev. Lett.*, **111**, 57002 (2013).

[29] V. E. Antonov, *J. Alloys Compd.*, **330–332**, 110 (2002).

[30] G. Wiesinger, and G. Hilscher, in Handbook of Magnetic Materials, Vol. 17, K. H. J. Buschow, (ed.) (North-Holland, Elsevier Science Publ. BV, 2008), pp. 293–456.

[31] L. George, and S. K. Saxena, *Int. J. Hydrogen Energy*, **35**, 5454 (2010).

[32] C. Donnerer, T. Scheler, and E. Gregoryanz, *J. Chem. Phys.*, **138**, 134507 (2013).

[33] M. S. Hasnain, C. Webb, and E. M. Gray, *Prog. Solid State Chem.*, **44**, 20 (2016).

[34] V. V. Struzhkin, *Physica C: Superconductivity Appl.*, **514**, 77 (2015).

[35] T. Bi, N. Zarifi, T. Terpstra, and E. Zurek, in *Reference Module in Chemistry, Molecular Sciences and Chemical Engineering* (Elsevier, 2019).

[36] N. W. Ashcroft, *Phys. Rev. Lett.*, **92**, 187002 (2004).

[37] M. Martinez-Canales, A. Bergara, J. Feng, and W. Grochala, *J. Phys. Chem. Solids*, **67**, 2095 (2006).

[38] J. S. Tse, Y. Yao, and K. Tanaka, *Phys. Rev. Lett.*, **98**, 117004 (2007).

[39] G. Gao, A. R. Oganov, A. Bergara, M. Martinez-Canales, T. Cui, T. Iitaka, Y. Ma, and G. Zou, *Phys. Rev. Lett.*, **101**, 107002 (2008).

[40] Y. Li, G. Gao, Y. Xie, Y. Ma, T. Cui, and G. Zou, *Proc. Natl. Acad. Sci. USA*, **107**, 15708 (2010).

[41] J. A. Flores-Livas, M. Amsler, T. J. Lenosky, L. Lehtovaara, S. Botti, M. A. L. Marques, and S. Goedecker, *Phys. Rev. Lett.*, **108**, 117004 (2012).

[42] H. Wang, J. S. Tse, K. Tanaka, T. Iitaka, and Y. Ma, *Proc. Natl. Acad. Sci. USA*, **109**, 6463 (2012).

[43] J. Hooper, B. Altintas, A. Shamp, and E. Zurek, *J. Phys. Chem. C*, **117**, 2982 (2013).

[44] Y. Li, J. Hao, H. Liu, Y. Li, and Y. Ma, *J. Chem. Phys.*, **140**, 174712 (2014).

[45] H. Liu, I. I. Naumov, R. Hoffmann, N. W. Ashcroft, and R. J. Hemley, *Proc. Natl. Acad. Sci. USA*, **114**, 6990 (2017).

[46] F. Peng, Y. Sun, C. J. Pickard, R. J. Needs, Q. Wu, and Y. Ma, *Phys. Rev. Lett.*, **119**, 107001 (2017).

[47] C. Heil, S. di Cataldo, G. B. Bachelet, and L. Boeri, *Phys. Rev. B*, **99**, 220502 (2019).

[48] A. P. Drozdov, M. I. Eremets, I. A. Troyan, V. Ksenofontov, and S. I. Shylin, *Nature*, **525**, 73 (2015).

[49] A. P. Drozdov, P. P. Kong, V. S. Minkov, S. P. Besedin, M. A. Kuzovnikov, S. Mozaffari, L. Balicas, F. F. Balakirev, D. E. Graf, V. B. Prakapenka, E. Greenberg, D. A. Knyazev, M. Tkacz, and M. I. Eremets, *Nature*, **569**, 528 (2019).

[50] M. Somayazulu, M. Ahart, A. K. Mishra, Z. M. Geballe, M. Baldini, Y. Meng, V. V. Struzhkin, and R. J. Hemley, *Phys. Rev. Lett.*, **122**, 027001 (2019).

[51] H. Mii, M. Senoo, and I. Fujishiro, *Jpn. J. Appl. Phys.*, **15**, 777 (1976).

[52] V. F. Degtyareva, G. V. Chipenko, I. T. Belash, O. I. Barkalov, and E. G. Ponyatovskii, *Phys. Status Solidi (A)*, **89**, K127 (1985).

[53] V. V. Brazhkin, V. V. Glushkov, S. V. Demishev, Y. V. Kosichkin, N. E. Sluchanko, and A. I. Shulgin, *J. Phys. Condens. Matter*, **5**, 5933 (1993).

[54] N. E. Sluchanko, V. V. Glushkov, S. V. Demishev, N. A. Samarin, A. K. Savchenko, J. Singleton, W. Hayes, V. V. Brazhkin, A. A. Gippius, and A. I. Shulgin, *Phys. Rev. B*, **51**, 1112 (1995).

[55] Y. Muraba, S. Matsuishi, S.-W. Kim, T. Atou, O. Fukunaga, and H. Hosono, *Phys. Rev. B*, **82**, 180512 (2010).

[56] M. Pfisterer, and G. Nagorsen, *Zeit. Naturforschung B*, **38**, 811 (1983).

[57] H. Kotegawa, H. Sugawara, and H. Tou, *J. Phys. Soc. Jpn.*, **78**, 013709 (2009).

[58] P. L. Alireza, Y. T. C. Ko, J. Gillett, C. M. Petrone, J. M. Cole, G. G. Lonzarich, and S. E. Sebastian, *J. Phys. Condens. Matter*, **21**, 012208 (2009).

[59] E. Colombier, S. L. Bud'ko, N. Ni, and P. C. Canfield, *Phys. Rev. B*, **79**, 224518 (2009).

[60] C. Gen-Fu, L. Zheng, L. Gang, H. Wan-Zheng, D. Jing, Z. Jun, Z. Xiao-Dong, Z. Ping, W. Nan-Lin, and L. Jian-Lin, *Chinese Phys. Lett.*, **25**, 3403 (2008).

[61] A. Leithe-Jasper, W. Schnelle, C. Geibel, and H. Rosner, *Phys. Rev. Lett.*, **101**, 207004 (2008).

[62] W. Schnelle, A. Leithe-Jasper, R. Gumeniuk, U. Burkhardt, D. Kasinathan, and H. Rosner, *Phys. Rev. B*, **79**, 214516 (2009).

[63] S. R. Saha, N. P. Butch, K. Kirshenbaum, and J. Paglione, *Phys. Rev. B*, **79**, 224519 (2009).

[64] F. Han, X. Zhu, P. Cheng, G. Mu, Y. Jia, L. Fang, Y. Wang, H. Luo, B. Zeng, B. Shen, L. Shan, C. Ren, and H.-H. Wen, *Phys. Rev. B*, **80**, 024506 (2009).

[65] H. L. Shi, H. X. Yang, H. F. Tian, J. B. Lu, Z. W. Wang, Y. B. Qin, Y. J. Song, and J. Q. Li, *J. Phys.: Condens. Matter*, **22**, 125702 (2010).

[66] Y. Nishikubo, S. Kakiya, M. Danura, K. Kudo, and M. Nohara, *J. Phys. Soc. Jpn.*, **79**, 095002 (2010).

[67] R. Cortes-Gil, and S. J. Clarke, *Chem. Mater.*, **23**, 1009 (2011).

[68] H. Hiramatsu, T. Katase, T. Kamiya, M. Hirano, and H. Hosono, *Phys. Rev. B*, **80**, 052501 (2009).

[69] S. R. Saha, N. P. Butch, T. Drye, J. Magill, S. Ziemak, K. Kirshenbaum, P. Y. Zavalij, J. W. Lynn, and J. Paglione, *Phys. Rev. B*, **85**, 024525 (2012).

[70] Z. Gao, Y. Qi, L. Wang, D. Wang, X. Zhang, C. Yao, C. Wang, and Y. Ma, *(Europhys. Lett.)*, **95**, 67002 (2011).

[71] B. Lv, L. Deng, M. Gooch, F. Wei, Y. Sun, J. K. Meen, Y.-Y. Xue, B. Lorenz, and C.-W. Chu, *Proc. Natl. Acad. Sci. USA*, **108**, 15705 (2011).

[72] G. Balducci, S. Brutti, A. Ciccioli, G. Gigli, P. Manfrinetti, A. Palenzona, M. Butman, and L. Kudin, *J. Phys. Chem. Solids*, **66**, 292 (2005).

[73] H. Okamoto, *J. Phase Equilibria Diffusion*, **27**, 428 (2006).

[74] V. Z. Turkevich, T. A. Prikhna, and A. V. Kozyrev, *High Pressure Res.*, **29**, 87 (2009).

[75] D. R. Lide (ed.), *CRC Handbook of Chemistry and Physics*, Vol. Internet Version (CRC Press, 2005).

[76] J. Karpinski, M. Angst, J. Jun, S. Kazakov, R. Puzniak, A. Wisniewski, J. Roos, H. Keller, A. Perucchi, L. Degiorgi, M. Eskildsen, P. Bordet, L. Vinnikov, and A. Mironov, *Supercond. Sci. Technol.*, **16**, 221 (2003).

[77] J. Karpinski, N. D. Zhigadlo, S. Katrych, R. Puzniak, K. Rogacki, and R. Gonnelli, *Physica C*, **456**, 3 (2007).

[78] S. Lee, *Physica C*, **456**, 14 (2007).

[79] D. Mou, R. Jiang, V. Taufour, R. Flint, S. L. Bud'ko, P. C. Canfield, J. S. Wen, Z. J. Xu, G. Gu, and A. Kaminski, *Phys. Rev. B*, **91**, 140502 (2015).

[80] D. Mou, R. Jiang, V. Taufour, S. L. Bud'ko, P. C. Canfield, and A. Kaminski, *Phys. Rev. B*, **91**, 214519 (2015).

[81] D. Mou, S. Manni, V. Taufour, Y. Wu, L. Huang, S. L. Bud'ko, P. C. Canfield, and A. Kaminski, *Phys. Rev. B*, **93**, 144504 (2016).

[82] H. Kim, K. Cho, M. A. Tanatar, V. Taufour, S. K. Kim, S. L. Bud'ko, P. C. Canfield, V. G. Kogan, and R. Prozorov, *Symmetry*, **11**, 1012 (2019).

[83] J. Nagamatsu, N. Nakagawa, T. Muranaka, Y. Zenitani, and J. Akimitsu, *Nature*, **410**, 63 (2001).

[84] K.-H. Jin, H. Huang, J.-W. Mei, Z. Liu, L.-K. Lim, and F. Liu, *NPJ Comp. Mater.*, **5**, 57 (2019).

[85] N. Zhigadlo, *J. Cryst. Growth*, **402**, 308 (2014).

[86] S. K. Kim, *Pressure Effects on Selected Correlated Electron Systems*, Ph.D. thesis, Iowa State University (2013).

Chapter 11

Future Directions in Quantum Materials Synthesis

T. M. McQueen[1], T. Berry[2], J. R. Chamorro[3], A. Ghasemi[4],
W. A. Phelan[*,†,5], E. A. Pogue[6], L. A. Pressley[7], M. Sinha[8],
V. J. Stewart[9], T. T. Tran[*,‡,10], H. K. Vivanco[11], and M. J. Winiarski[*,12]

*Department of Chemistry, Department of Physics and Astronomy,
Department of Materials Science and Engineering,
Institute for Quantum Matter, and the Platform for the
Accelerated Realization, Analysis, and Discovery of Interface Materials,
The Johns Hopkins University, Baltimore, MD 21218, USA
*Present Address: Department of Solid State Physics, Faculty of
Technical Physics and Applied Mathematics, Gdańsk University of
Technology, Gdańsk, Poland
†Present Address: Los Alamos National Laboratory,
New Mexico 87544, USA
‡Present Address: Department of Chemistry, Clemson University,
Clemson, SC 29634, USA*
[1] mcqueen@jhu.edu
[2] tberry9@jhu.edu
[3] jchamor1@jhu.edu
[4] aghasem2@jhu.edu
[5] wap@lanl.gov
[6] epogue1@jhu.edu
[7] lpressl3@jhu.edu
[8] msinha4@jhu.edu
[9] vstewar4@jhu.edu
[10] thao@clemson.edu
[11] hvivanc1@jhu.edu
[12] michal.winiarski@pg.edu.pl

11.1. Introduction

In this book, there have been many examples of the state of the art in quantum materials. Many more exciting capabilities are emerging, from combined MBE-MOMBE/CVD-ARPES [1] and ultra-high pressure optical floating zone synthesis [2, 87] through the NSF Materials Innovation Platforms and investments by private foundations, to new theoretical spectroscopy methods [35]. The purpose of this chapter is to be forward looking and address key questions: What will the quantum materials of the future be? How will we identify, synthesize, and harness them?

Of course to do so, it is first valuable to consider in more detail existing classes of quantum materials and their already known, or clearly foreseen, uses, Figure 11.1 [21–25]. Already, quantum materials have passed into technological use, with the widespread application of superconductors in medicine and research, and the early development of quantum computers [85, 86]. Applications range from using superconductors to generate the large magnetic fields needed for spectroscopy and imaging (e.g., in NMRs and MRIs) to the use of thin film ferroelectrics in computer memory devices (e.g.,

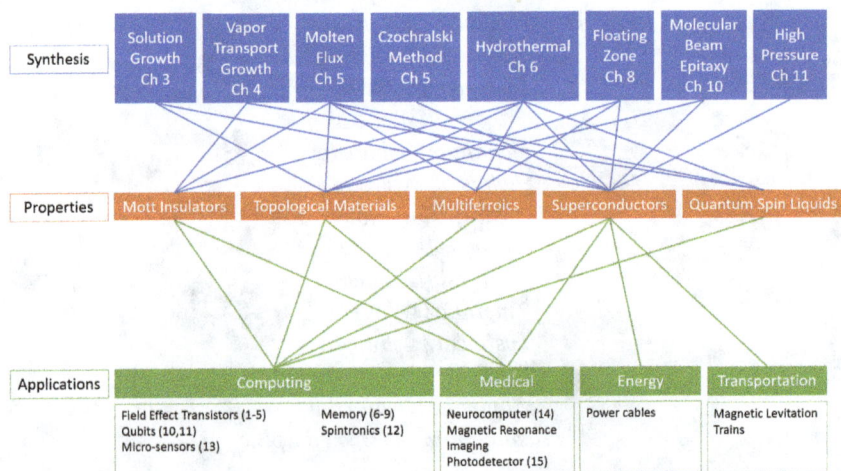

Figure 11.1.   Already known applications of common classes of quantum materials [6–20].

FeRAM) and the use of quantized electron states in nanoparticles for high brightness and contrast display screens.

As the synthesis and discovery of quantum materials improves, these technological applications will greatly expand. Topological materials may be applied in quantum computing or information by making use of their spintronic properties, their dissipationless current flow along edge states, or the use of Majorana fermions to stabilize qubits [87]. High-temperature superconductors integrated into power grids could dramatically reduce energy expenditures. Mott metal-insulator transitions may be used in improved transistors and memory devices, or magnetoelectric materials in memory devices and sensors [87, 88].

The future directions for quantum materials research discussed here — the application of big data to discovery, enhanced synthetic control, and improved understanding of defects — heighten the prospect of further development of these technologies.

11.2. The Current State of the Art

Before exploring the future, we must consider the present state of the art approaches. The fusion of synthetic and characterization advances with computational and theoretical tools is driving the current decade of quantum materials research. Figure 11.2 shows a schematic view of forefront methods of synthesis and characterization of quantum materials. The major directions of development in the field of synthesis are: (i) engineering of materials at atomic scales, including control and engineering of defects [26–28] and various epitaxial growth methods [29], (ii) computer-assisted solid-state chemistry — modeling of thermodynamic landscape of reactions, high throughput calculations and relational materials databases, such as Materials Project, Materiae, or OQMD, that offer a broad view of the crystallographic and physical data and allow the user to draw general predictions [30–34], and advanced theoretical methods that are able to predict the physical many-body properties from first principles ("theoretical spectroscopy") [35], (iii) chemical control over the synthesis, e.g., by establishing a kinetic control or employing

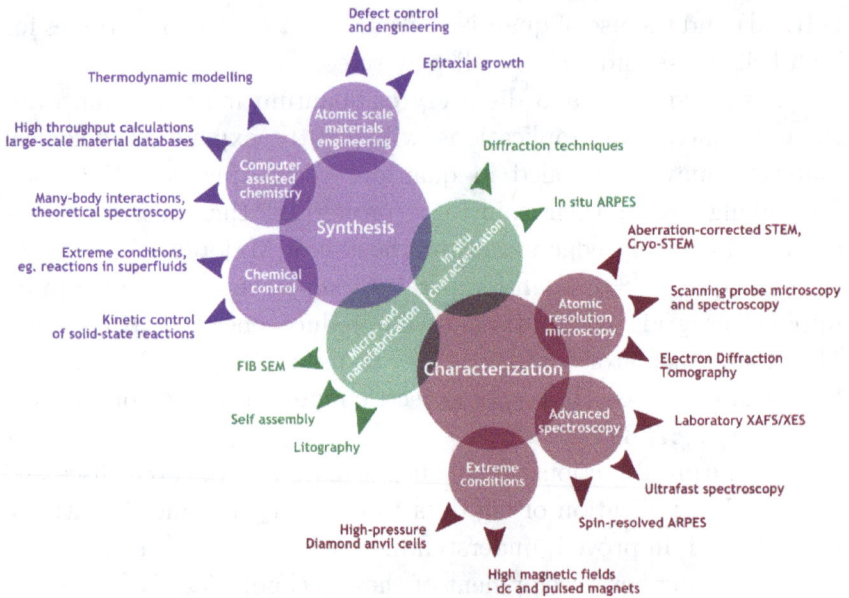

Figure 11.2.   Current common synthetic and characterization tools and frontiers in quantum materials discovery [26–51].

extreme reaction conditions to yield the desired product [36–38]. An example of how these concepts translate into actual practice is shown in Figure 11.3, a modern workflow for the development of materials in single-crystalline form (one of many forms useful for fundamental and applied studies).

In the field of characterization, the three major directions are: (i) atomic resolution microscopy and spectroscopy, especially useful in case for studying reduced dimensionality systems and defect control [39–43], (ii) advanced spectroscopic techniques, such as ultrafast spectroscopy [44–46], spin-resolved ARPES [47], or laboratory XAS/XES [48], that provide details on the electronic structure of materials, and (iii) extreme conditions e.g., very high magnetic fields.

Finally, micro- and nanofabrication methods enable not only the production of metamaterials and nano-devices, but also are useful

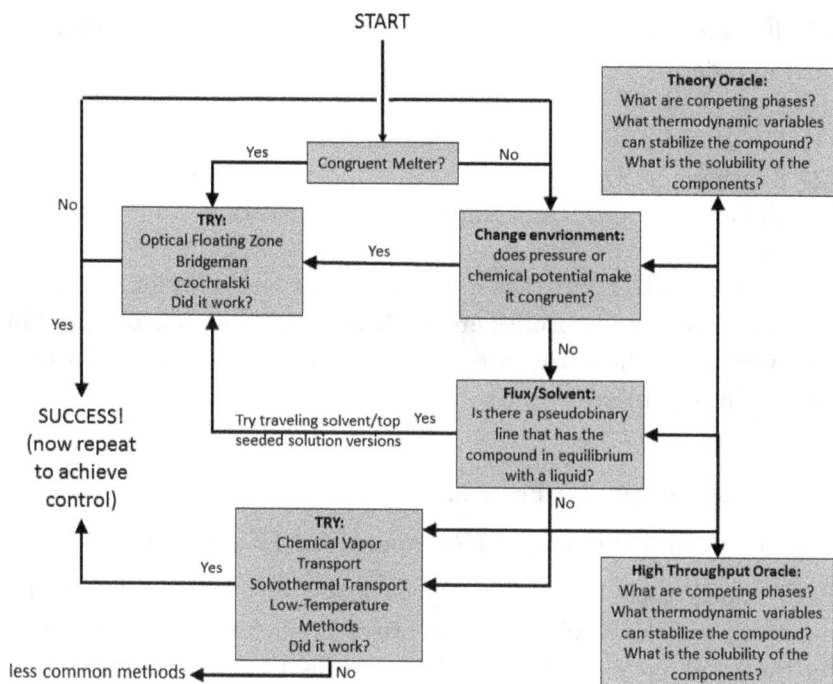

Figure 11.3. Current state of the art workflow to prepare single crystals of quantum materials.

in studying the detailed physical properties of microscale systems [49]. *In-situ* characterization methods in turn are useful for studying the reaction pathways [38, 50, 51], as well as for characterization of materials without exposing them to the environment which may affect their properties.

11.3. The Frontiers

Given the current state of the art in quantum materials synthesis, it is clear that substantial progress has been made over the past 50 years. At the same time, comparison to other fields — including astronomy, organic chemistry, and biology — shows that there are numerous opportunities to make further significant advances in the synthesis of quantum materials. Loosely, we identify three areas in which there is substantial opportunity to make progress within the next decade:

(1) Computer-Aided Observational Discovery,
(2) Synthesis by Design,
(3) Elaboration and Harnessing of Tuned Defects.

Each of these frontiers is in fact a specific instantiation of a more general principle: discoveries are made by looking in new places or in new ways, and these are three accessible materials handle in which to look in new places.

11.4. Exploring the Frontiers

11.4.1. *Computer-aided Observational Discovery*

Right now, there are two conventional processes for advancing quantum materials: either (i) theorists predict materials that can be the candidates for the novel properties that are interesting to study, followed by experimentalists synthesizing materials to realize these models, or (ii) the inverse in which unexpected phenomena arise in experiments and lead to the development of new theories. Both are instantiations of the well-known materials by design loop. Given the prevalent use of automation for comparable processes in organic chemistry and biology, it is surprising that there is comparatively little automated assistance for synthesis in quantum materials.

There is a good reason for this. Quantum materials discovery is driven by making unexpected observations, and traditional automation workflow, while excellent for optimization, necessarily decreases human observations of the data. This is an exceptionally challenging computer science problem, but recent demonstrations of computer driven discovery, e.g., identification of new reaction pathways [3], suggest that it may be possible to automate quantum materials synthesis while preserving that critical aspect of discovery — the recognition of, and ability to act upon, unexpected observations.

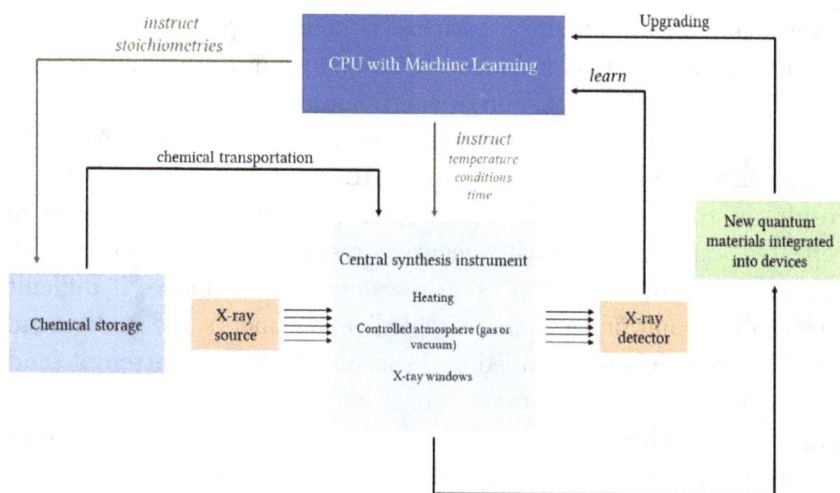

Figure 11.4. Automating true quantum materials discovery as has been done in organic chemistry [3] and drug discovery [4] is a frontier.

It is alluring to think that analogous systems can be built for quantum materials, one schematic of which is laid out in Figure 11.4. Which ingredients do we already possess, and what ingredients do we need, to make this a reality?

Many of the existing efforts to compile with materials data and use them for material discovery have been carried out in the computational realm. As part of or inspired by the Materials Genome Initiative and the EU Horizon 2020 program, a variety of databases including NOMAD (Novel Materials Discovery) [52], OQMD (Open Quantum Materials Database) [33], AiiDA (Automated Interactive Infrastructure and Database for Computational Science) [54], the Computational Materials Repository [55], AFLOW [56, 57], and Materials Project [58] have been created. Many of these databases are sortable by chemistry (elements).

Once a material has been theorized, it must be synthesized to be of practical use. Its properties must also be tested to ensure that the theory is correct. These properties in real materials are affected by the processing, of the material, including the synthesis process. These final two aspects of accelerated materials discovery have, to date, been much more difficult to address. Because of processing-induced variation, chemistry, alone, is an incomplete (and often misleading) indexing scheme for cataloging experimental data.

One major issue in obtaining useful datasets for Artificial Intelligence/Machine Learning (AI/ML) is the Lack of negative results in the published literature [59]. In addition to resulting in repeated experiments with "negative results" by different groups, the lack of reported "failed" syntheses and growth makes it difficult to glean information from these "failed" experiments and understand the difference between a failed and successful synthesis attempt (and the range of conditions under which success is possible). Machine learning algorithms require data related to these failed experiments as much as records of successful experiments to allow for the calculation of these regions of synthesis success [59]. Unfortunately, databases of negative results, when they exist [59], tend to only contain submissions from the originating group, likely because academic and commercial competition discourage the sharing of this information [54]. For AI and machine learning to reach their full potential in the materials realm, this human barrier must be overcome. Nonetheless, a variety of organizations have been addressing this issue of the lack of published negative results and inaccessible data. As part of the Materials Genome Initiative, NIST created a materials data curation system called MDCS, which is available in the NIST GitHub repository [60].

Practically, a major limitation to automation is the proliferation of incompatible equipment, particularly in university research environments. Non-computerized equipment for material synthesis and growth is common and computerized equipment is often more expensive and less customizable than home-built equipment (and too expensive for many institutions). Even for equipment that is computerized, incompatible or old unsupported operating systems make the integration of this equipment difficult. Integrating some characterization techniques with synthesis processes can be difficult given the different timescales involved. Some tasks are more amenable to automation than others. The creation of shared facilities may help a broader range of researchers access these tools and allow for such integrated approaches, although a variety of questions remain. According to the Clean Energy Materials Innovation Challenge Expert Workshop, the major questions such an approach would bring include: Where should such facilities be located? How does one access these facilities? How is the optimum use of the individual growth and characterization chambers managed? and How to manage intellectual property [61]?

Another concern with enhanced materials discovery is whether these techniques deepen our understanding of how chemistry and the natural world work. If these systems are used only as optimizers and are not used to probe the science behind reactions, revolutionary discovery may be stifled. On the other hand, higher throughputs may also enable better science, with more time to devote to analyzing trends and developing theory than time-consuming lab work. To do this, machine learning must give human-understandable explanations, inferences, and conclusions and must be able to autonomously reason given prior data [61]. Furthermore, a machine (that includes humans) is only as good as its sensors. If humans set the "goals" for the machine but do not see the nature of the "failed" samples, the utility of those "failed" materials may not be investigated (a goal can change). This may not be noticed if the AI cannot sense that this "failed" material is of interest, too.

To our knowledge, there are currently three autonomous robotic systems doing useful scientific research without human intervention (feeding their results back into their experimental models). Adam [62, 63], and Eve [63], which were designed by researchers at Aberystwyth University and Cambridge University, focus on biological research while the ARES robot, which is was created by the United States Air Force Research Laboratory, focuses on materials research and combines iterative experiments with *in-situ* materials characterization [64]. ARES autonomously learned to grow single-walled carbon nanotubes at well-defined rates, optimizing across multiple experimental parameters simultaneously. Further advances in AI and ML should allow for the automation of more material synthesis and characterization tasks, eventually enabling large, integrated systems. Some of these advances may also accelerate the research-to-industry pipeline, since a high-throughput goal is considered earlier in the process and potentially, a wider parameter space can be evaluated earlier in the development process. The ability to mine old data from "unsuccessful" syntheses can further accelerate that pipeline.

We note in passing that such efforts to automate crystal growth and synthesis processes and share data may enable both the growth of new materials and the wider adoption of these materials by a broader community in new technologies. But, in doing so, we must also maintain the ability to make use of the data that we have, ensuring that it is in accessible formats that are easily searched — this is essential for taking full advantage of recent advances in data analytics (currently popularized as "AI/Machine Learning"), as well as for enabling smaller scientific centers and "less-developed" countries to contribute more to the research on quantum materials.

11.4.2. *Synthesis by Design*[a]

Figure 11.5 shows main ways to control synthesis by design, as summarized in an excellent recent review [65]. "Materials by design" has become a forefront of the solid-state field and has had a direct impact on society, but the question remains: how do we truly achieve synthetic control?

Unlike fields such as organic chemistry, which has a known toolbox of reactions that facilitate the synthesis of molecules through functional group transformations, solid-state chemistry does not have a universal set of reactions for the synthesis of a desired product — nor is there likely to be one, given the sheer amount of materials possible with around 100 different usable elements. Instead, solid-state chemists can aim to create universal synthetic techniques involved so that they may be understood completely, and through their understanding discover new materials and phenomena.

Regarding metastable materials, their synthesis may be broken down across all synthetic techniques (universalization) into two separate categories: either we modify reaction dynamics to make

---

[a]We thank John Mitchell, Argonne National Laboratory, for this terminology.

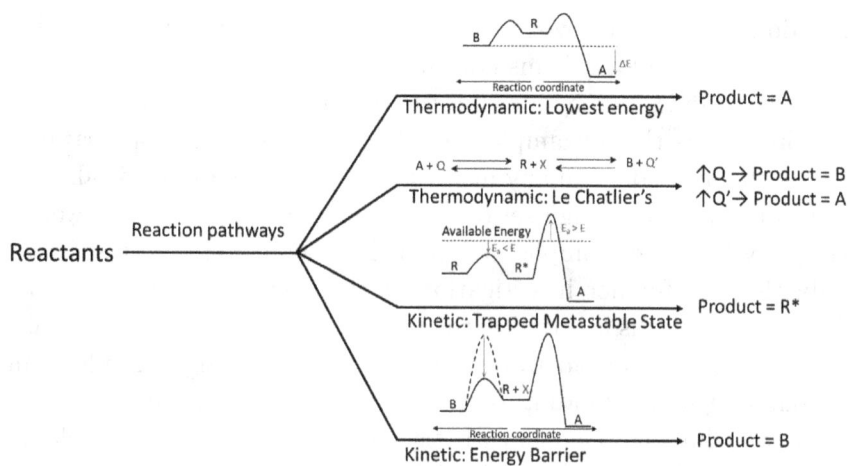

Figure 11.5. Controlling the reactivity landscape — and the rearrangement of atoms and the bonds between them — is a frontier to complete the materials by design loop to enable true synthesis by design.

them the most stable outcome of their synthesis reaction, or we shift reaction equilibria to favor their formation despite their instability. An example of the former is the synthesis of the metastable battery material VS_2, which is generated from the oxidative deintercalation of Li^+ from $LiVS_2$ with iodine via the reaction [82]:

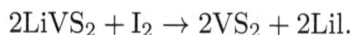

$$2LiVS_2 + I_2 \rightarrow 2VS_2 + 2LiI.$$

The formation of highly thermodynamically stable LiI byproduct modifies the overall reaction dynamics to favor the formation of metastable VS_2 despite its instability and inability to form via other conventional synthesis techniques. An example of the latter is the removal of Zn^{2+} from valence bond solid $LiZn_2Mo_3O_8$ via the reaction [83, 84]:

$$LinZn_2Mo_3O_8 + xI_2 \rightarrow LiZn_{2-x}Mo_3O_8 + xZnI_2.$$

When reacted with iodine, one would expect the removal of Li^+ from the lattice due to the high stability of LiI compared to ZnI_2. However, due to the volatility of ZnI_2, the formation of this phase and subsequent removal from the reaction through volatization shifts the reaction equilibrium towards the product side. Thus, the metastable $LiZn_{2-x}Mo_3O_8$ phase forms not due to a change in the total reaction thermodynamics but by altering the reaction equilibrium.

Specific as these examples may be, they present an opportunity to ask the question: can any material be synthesized if one is allowed to sample a large enough set of synthesis parameters? The answer to this may be rather complicated, and shall be answered in the future only through further investigation of synthesis techniques and the actual mechanisms involved in the formation of metastable materials.

So, what lies ahead to achieve synthesis by design? Aside from a complete understanding of current techniques, the development of new technologies and synthesis methods will push the field of quantum materials forward. In the short term, a shift of focus from trying to understand the materials themselves during synthesis and instead changing their environment may help bolster discovery. For example, the development of novel high-pressure systems capable of transforming gaseous environments into supercritical fluids may present a new forefront in discovery. Supercritical fluids, aside from acting as high pressure mediums, may also act as unconventional solvents that facilitate solid-state synthesis by decreasing the diffusive barrier to reaction. Another example involves changing the chemical potential of a synthesis environment, such that the interaction between the sample and its environment becomes nontrivial and thus reaction dynamics are altered. Looking further ahead, a truly complete understanding of materials synthesis by design will require a complete tracking, understanding, and control over all variables that often make synthesis possible for one group but irreproducible for another. A movement towards a universal understanding of solid-state synthesis is a movement towards complete reaction control.

11.4.3. *Elaboration and Harnessing of Tuned Defects*

A frontier closely related to synthesis by design is to recognize the extreme importance of defects — especially those beyond point defects — in governing the underlying physics. One of the main differences between theory and experiment is the fact that, for theory, we consider the system ideal and perfect to simplify the problem and make models. These assumptions help to predict interactions and their collective effects which cause the emergence of exotic phases. However, in experiments, the collected data is always messier than what theory predicts, sometimes to the extent that the expected features cannot be seen. The main reason is that the materials are not perfect. Disorders and imperfections are essentially unavoidable in solid-state materials. Their effects are often described in a negative context. Andrew Mackenzie summarized the general feeling toward disorders in the statement: "For the most part, disorders in condensed matter is a pain in the neck and a barrier to truth and enlightenment."

However, it is important to remember that defects and disorders that increase the entropy are also ways that system reduces its classical energy. Therefore, their effect should be considered in a Hamiltonian that describes the interactions of a system. There are two approaches to studying the defects (1) What introduces them to the structure and how they can be minimized? (2) How they influence the properties?

All real systems contain various types of imperfections which can be divided based on the dimensionality. Although defects are inevitable in the structure, they can be minimized by improving the with synthetic method. The nature of the defects and the number of defects are the first clues for determining how they are introduced to the structure. For example, the existence of cracks in a single crystal can be a sign that the material undergoes thermal stresses during the single-crystal growth [66, 67], and secondary phases can be a sign of coprecipitation or competitive reactions [68] (Figure 11.6).

The importance of studying the contribution of defects to the physical properties of materials has only increased over the past two decades. However, the idea of including the effect of defects in Hamiltonian goes back almost a century: Landau [69, 70] came with an approximate quantitative description for phase transitions and a general framework for classifying the phases in macroscopic many-body systems. He used the expansion of the free energy density. Building upon that work, Imry and Ma [71] modeled the contribution of detects by comparing the effect of symmetry breaking defects to

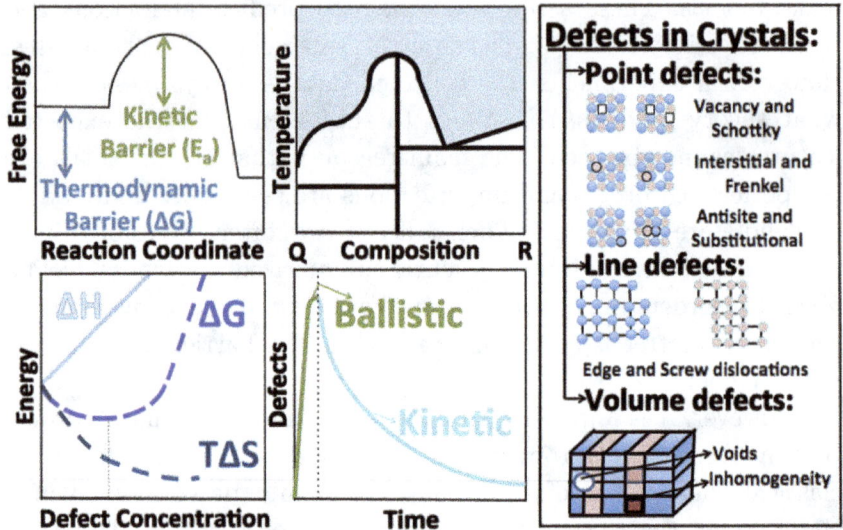

Figure 11.6.   In everything from structural metals to electronic semiconductors, "defects rule," and control the underlying physics. The same is assuredly true in quantum materials, and it is a frontier to explore how defects (particularly extended defects) control quantum phenomena in materials.

a random field. Currently significant research efforts focus on phases that do not follow landau's classification but feature unconventional; topological order. This order is due to the long-range entanglement of their quantum wave function [72–77]. The combination of introducing defects and disorders in the Hamiltonian and understanding how to control the defects in the structure let us tune the defects especially in the cases in which defects can induce the features [77–81].

Put more bluntly, all real materials have defects. A frontier in quantum materials is to learn how to control the position, distribution, and type of defects, and how to use them to measure (and, eventually, harness) the underlying quantum phenomenon. The tools for full characterization of defects from the atomic to microscale exist thanks to advances in the last 30 years in materials science — it is long past time these be applied to quantum condensed matter systems and forever break down the simple assumption that a crystalline material is periodic. We note in passing one recent theoretical attempt to do this has postulated the ability to metallize

Figure 11.7. A future grocery list? Probably not — and that means quantum materials need to adapt to not require cryogenic conditions.

a quantum spin liquid due to non-epitaxial proximity to a Dirac semimetal [88] (Figure 11.7).

11.5. Conclusion

The rapid developments in the field of quantum materials, such as quantum computing, beg the question: "when can I have those technologies in my hands? Am I going to have a quantum computer at my home? Do I need to add liquid helium in my grocery list?" When it comes to answering these questions we should remember that accessibility of those technologies do not necessarily mean that we are going to have it at our homes or in our hands. Rather, there can be facilities with remote access which holds the quantum system in proper physical conditions. However, even to reach that point the current methods should be improved and become more robust to have the highest possible efficiency.

In this chapter, we have reviewed the current state of the art in quantum materials synthesis, and identified some key frontiers

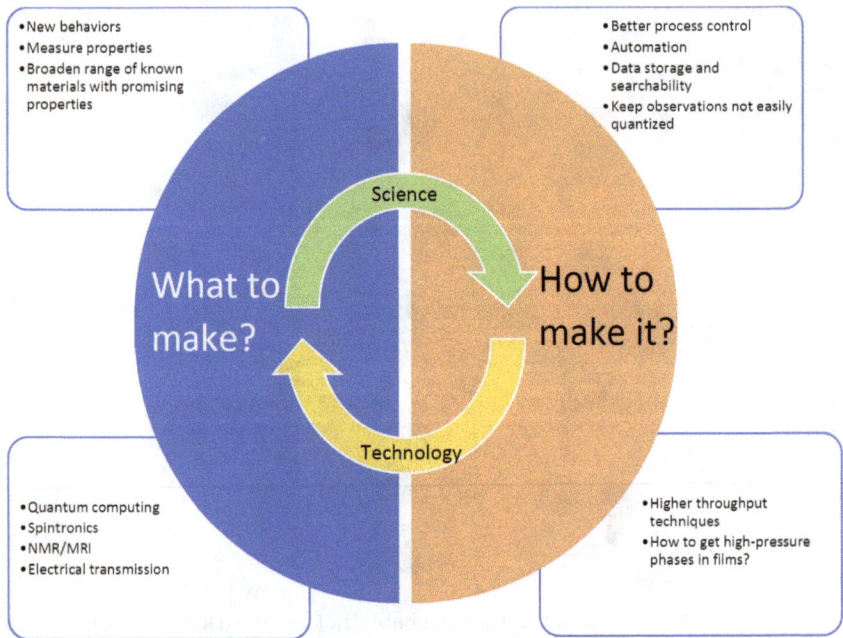

Figure 11.8. Tight integration across disciplines is critical to push quantum materials synthesis and discovery.

that deserve immediate pursuit, driven by a tighter coupling between scientific disciplines and advances in computing and related areas, Figure 11.8.

Bibliography

[1] D. G. Schlom, and K. M. Shen, *Appl. Surface Sci.* in Press (2019).

[2] W. A. Phelan, J. Zahn, Z. Kennedy, and T. M. McQueen, *J. Sol. St. Chem.*, **270**, 705 (2019).

[3] J. M. Granda, L. Donina, V. Dragone, D.-L. Long, and L. Cronin, *Nature*, **559**, 377 (2018).

[4] B. Chen, and A. J. Butte, *Big Data*, **99**, 285 (2015).

[5] A. Pergament, A. Crunteanu, and A. Beaumont, arxiv:1601.06246 (2015).

[6] Y. Zhou, and S. Ramanathan, *Critical Rev. Solid State Mater. Sci.*, **38**, 286 (2013).

[7] J. Son, S. Rajan, S. Stemmer, and S. Allen, *J. Appl. Phys.*, **11**, 8 (2011).

[8] R. Waser, and M. Aono, *Nat. Mater.*, **6**, 833 (2007).

[9] A. Sawa, *Mater. Today.*, **11**, 28 (2008).

[10] V. E. Wood, and A. E. Austin, in *Magnetoelectric Interaction Phenomena in Crystals*, A. J. H. Schmid (ed.) (Gordon and Breach, 1975).

[11] N. Fujimura, T. Ishida, T. Yoshimura, and T. Ito, *Appl. Phys. Lett.*, **69**, 1011 (1996).

[12] R. Ramesh, and N. A. Spaldin, *Nature Mater.*, **6**, 21 (2007).

[13] H. N. Lee, Y. T. Kim, and Y. K. Park. *Appl. Phys. Lett.*, **74**, 3887 (1999).

[14] J. Gambetta, J. Chow, and M. Steffen, *Quantum Inform.*, **3**, 2 (2017).

[15] X. Zhang, H.-O. Li, K. Wang, G. Cao, M. Xaio, and G.-P. Guo, *Chinese Phys. B*, **27**, 2 (2018).

[16] B. H. Bairamov *et al.*, in A. Katashev, Y. Dekhtyar, and J. Spigulis (eds), *14th Nordic-Baltic Conference on Biomedical Engineering and Medical Physics*, IFMBE Proceedings, Vol 20 (Springer, 2008).

[17] Y. Yan *et al.*, *Nano Lett.*, **14**, 4389 (2014).

[18] J. Chang, L. F. Register, and S. K. Banerjee, *J. Appl. Phys.*, **112**, 3045 (2012).

[19] W. G. Vandenberghe, and M. V. Fischetti, in *Proc. 2014 IEEE Int. Electron Devices Meeting*, San Francisco, CA, USA (15–17 December 2014).

[20] https://science.osti.gov/-/media/bes/pdf/reports/2016/BRNQM_rpt_Final_12-09-2016.pdf

[21] https://science.osti.gov/-/media/bes/pdf/reports/2018/Quantum_systems.pdf

[22] https://www.nsf.gov/mps/dmr/MIQM_report_v15.pdf

[23] https://www.nae.edu/204196/Frontiers-of-Materials-Research

[24] https://www.nap.edu/read/12640/chapter/1

[25] R. R. Nair, M. Sepioni, I.-L. Tsai, O. Lehtinen, J. Keinonen, A. V. Krasheninnikov, T. Thomson, A. K. Geim, and I.V. Grigorieva, *Nat. Phys.*, **8**, 199 (2012).

[26] J. Cervenka, M. I. Katsnelson, and C. F. J. Flipse, *Nat. Phys.*, **5**, 840 (2009).

[27] K. E. Arpino, B. A. Trump, A. O. Scheie, T. M. McQueen, and S. M. Koohpayeh, *Phys. Rev. B*, **95**, 094407 (2017).

[28] T. P. Ginley, Y. Wang, and S. Law, *Crystals*, **6**, 154 (2016).

[29] Y. Xu, M. Yamazaki, and P. Villars, *Japanese J. Appl. Phys.*, **50**, 11RH02 (2011).

[30] T. G. Mitina, and V. A. Blatov, *Crystal Growth Design*, **13**, 1655 (2013).

[31] A. Jain, S. P. Ong, G. Hautier, W. Chen, W. D. Richards, S. Dacek, S. Cholia, D. Gunter, D. Skinner, and G. Ceder, *APL Mater.*, **1**, 011002 (2013).

[32] S. Kirklin, J. E. Saal, B. Meredig, A. Thompson, J. W. Doak, M. Aykol, S. Ruhl, and C. Wolverton, *NPJ Comput. Mater.*, **1**, 15010 (2015).

[33] T. Zhang, Y. Jiang, Z. Song, H. Huang, Y. He, Z. Fang, H. Weng, and C. Fang, arXiv:1807.08756 (2018).

[34] P. R. C. Kent, and G. Kotliar, *Science*, **361**, 348 (2018).

[35] M. G. Kanatzidis, K. R. Poeppelmeier, S. Bobev, A. M. Guloy, S.-J. Hwu, A. Lachgar, S. E. Latturner, E. Schaak, D.-K. Seo, and S. C. Sevov, *Prog. Solid State Chem.*, **36**, 1 (2008).

[36] L. Soderholm, and J. F. Mitchell, *APL Materials*, **4**, 053212 (2016).

[37] A. J. Martinolich, and J. R. Neilson, *Chem. Mater.*, **29**, 479 (2017).

[38] C. F. Hirjibehedin, and Y. Wang, *J. Phys. Condens. Matter*, **26**, 390301 (2014).

[39] N. Tanaka, *Sci. Technol. Adv. Mater.*, **9**, 014111 (2008).

[40] B. H. Savitzky, I. El Baggari, C. B. Clement, E. Waite, B. H. Goodge, D. J. Baek, J. P. Sheckelton, C. Pasco, H. Nair, N. J. Schreiber, J. Hoffman, A. S. Admasu, J. Kim, S.-W. Cheong, A. Bhattacharya, D. G. Schlom, T. M. McQueen, R. Hovden, and L. F. Kourkoutis, *Ultramicroscopy*, **191**, 56 (2018).

[41] U. Kolb, E. Mugnaioli, and T. E. Gorelik, *Crystal Res. Technol.*, **46**, 542 (2011).

[42] T. E. Gorelik, A. A. Stewart, and U. Kolb, *J. Microscopy*, **244**, 325 (2011).

[43] R. D. Averitt, and A. J. Taylor, *J. Phys. Condens. Matter*, **14**, R1357 (2002).

[44] J. Orenstein, *Phys. Today*, **65**, 44 (2012).

[45] C. Giannetti, M. Capone, D. Fausti, M. Fabrizio, F. Parmigiani, and D. Mihailovic, *Adva. Phys.*, **65**, 58 (2016).

[46] C. Bigi, P. K. Das, D. Benedetti, F. Salvador, D. Krizmancic, R. Sergo, A. Martin, G. Panaccione, G. Rossi, J. Fujii, and I. Vobornik, *J. Synchrotron Rad.*, **24**, 750 (2017).

[47] E. P. Jahrman, W. M. Holden, A. S. Ditter, D. R. Mortensen, G. T. Seidler, T. T. Fister, S. A. Kozimor, L. F. J. Piper, J. Rana, N. C. Hyatt, and M. C. Stennett, arXiv:1807.08059 (2018).

[48] J. T. Mlack, A. Rahman, G. Danda, N. Drichko, S. Friedensen, M. Drndic, and N. Markovic, *ACS Nano*, **11**, 5873 (2017).

[49] D. P. Shoemaker, Y.-J. Hu, D. Y. Chung, G. J. Halder, P. J. Chupas, L. Soderholm, J. F. Mitchell, and M. G. Kanatzidis, *Proc. Nat. Acad. Sci.*, **111**, 10922 (2014).

[50] D. Abeysinghe, A. Huq, J. Yeon, M. D. Smith, and H.-C. zur Loye, *Chem. Mater.*, **30**, 1187 (2018).

[51] NOMAD Repos. Nomad-Repository.Eu (2018).

[52] G. Pizzi, A. Cepellotti, R. Sabatini, N. Marzari, and B. Kozinsky, *Comput. Mater. Sci.*, **111**, 218 (2016).

[53] D. D. Landis, J. S. Hummelshoj, S. Nestorov, J. Greeley, M. Dulak, T. Bligaard, J. K. Norskov, and K. W. Jacobsen, *Comput. Sci. Eng.*, **14**, 51 (2012).

[54] S. Curtarolo, W. Setyawan, G. L. W. Hart, M. Jahnatek, R. V. Chepulskii, R. H. Taylor, S. Wang, J. Xue, K. Yang, O. Levy, M.J. Mehl, H. T. Stokes, D. O. Demchenko, and D. Morgan, *Comput. Mater. Sci.*, **58**, 218 (2012).

[55] S. Curtarolo, W. Setyawan, S. Wang, J. Xue, K. Yang, R. H. Taylor, L. J. Nelson, G. L. W. Hart, S. Sanvito, M. Buongiorno-Nardelli, N. Mingo, and O. Levy, *Comput. Mater. Sci.*, **58**, 227 (2012).

[56] A. Jain, S. P. Ong, G. Hautier, W. Chen, W. D. Richards, S. Dacek, S. Cholia, D. Gunter, D. Skinner, G. Ceder, and K. A. Persson, *APL Mater.*, **1**, 011002 (2013).

[57] P. Raccuglia, K. C. Elbert, P. D. F. Adler, C. Falk, M. B. Wenny, A. Mollo, M. Zeller, S. A. Friedler, J. Schrier, and A. J. Norquist, *Nature*, **533**, 73 (2016).

[58] A. Dima, S. Bhaskarla, C. Becker, M. Brady, C. Campbell, P. Dessauw, R. Hanisch, U. Kattner, K. Kroenlein, M. Newrock, A. Peskin, R. Plante, S.-Y. Li, P.-F. Rigodiat, G.S. Amaral, Z. Trautt, X. Schmitt, J. Warren, and S. Youssef, *JOM*, **68**, 2053 (2016).

[59] A. Aspuru-Guzik, K. Persson, and H. Tribukait-Vasconcelos, *Materials Acceleration Platform-Accelerating Advanced Energy Materials Discovery by Integrating High-Throughput Methods with Artificial Intelligence* (Mexico City, Mexico, 2017).

[60] R. D. King, J. Rowland, S. G. Oliver, M. Young, W. Aubrey, E. Byrne, M. Liakata, M. Markham, P. Pir, L. N. Soldatova, A. Sparkes, K. E. Whelan, and A. Clare, *Science*, **324**, 85 (2009).

[61] A. Sparkes, W. Aubrey, E. Byrne, A. Clare, M. N. Khan, M. Liakata, M. Markham, J. Rowland, L. N. Soldatova, K. E. Whelan, M. Young, and R. D. King, *Autom. Exp.*, **2**, 1 (2010).

[62] P. Nikolaev, D. Hooper, F. Webber, R. Rao, K. Decker, M. Krein, J. Poleski, R. Barto, and B. Maruyama, *NPJ Comput. Mater.*, **2**, 16031 (2016).

[63] J. R. Chamorro, and T. M. McQueen, *Acc. Chem. Res.*, **51**, 2918 (2018).

[64] S. M. Koohpayeh, D. Fort, and J. S. Abell, *Prog. Cryst. Growth Charact. Mater.*, **54**, 121 (2008).

[65] S. M. Koohpayeh, *Prog. Cryst. Growth Charact. Mater.*, **62**, 22 (2016).

[66] W. A. Phelan, S. M. Koohpayeh, P. Cottingham, J. A. Tutmaher, J. C. Leiner, M. D. Lumsden, C. M. Lavelle, X. P. Wang, C. Hoffmann, M. A. Siegler, N. Haldolaarachchige, D. P. Young, and T. M. McQueen, *Sci. Rep.*, **6**, 1 (2016).

[67] L. D. Landau, *Zh. Eks. Teor. Fiz.*, **7**, 19 (1937).

[68] V. L. Pokrovsky, *Phys.-Uspekhi*, **52**, 1169 (2009).

[69] Y. Imry, and S. K. Ma, *Phys. Rev. Lett.*, **35**, 1399 (1975).

[70] F. Alet, A. M. Walczak, and M. P. A. Fisher, *Phys. A Stat. Mech. Appl.*, **369**, 122 (2006).

[71] C. Castelnovo, S. Trebst, and M. Troyer, Topological order and quantum criticality, (2009).

[72] C. Castelnovo, R. Moessner, and S. L. Sondhi, Spin ice, fractionalization and topological order, *Annu. Rev. Condens. Matter Phys.*, **3**, 35 (2012).

[73] X.-G. Wen, Topological order: From long-range entangled quantum matter to an unification of light and electrons, *ISRN Condensed Matter Physics*, **2013**, 198710 (2013).

[74] T. Senthil, Symmetry-Protected Topological Phases of Quantum Matter, *Annu. Rev. Condens. Matter Phys.*, **6**, 299 (2015).

[75] L. Savary, and L. Balents, *Phys. Rev. Lett.*, **118**, 1 (2017).

[76] C. Z. Chen, J. Song, H. Jiang, Q. F. Sun, Z. Wang, and X. C. Xie, *Phys. Rev. Lett.*, **115**, 1 (2015).

[77] S. Seo, X. Lu, J. X. Zhu, R. R. Urbano, N. Curro, E. D. Bauer, V. A. Sidorov, L. D. Pham, T. Park, Z. Fisk, and J. D. Thompson, *Nat. Phys.*, **10**, 120 (2014).

[78] B. C. Camargo, Y. Kopelevich, A. Usher, and S. B. Hubbard, *Appl. Phys. Lett.*, **108**, 3 (2016).

[79] J. J. Wen, S. M. Koohpayeh, K. A. Ross, B. A. Trump, T. M. McQueen, K. Kimura, S. Nakatsuji, Y. Qiu, D. M. Pajerowski, J. R. D. Copley, and C. L. Broholm, *Phys. Rev. Lett.*, **118**, 1 (2017).

[80] D. W. Murphy, C. Cros, F. J. DiSalvo, and J. V. Waszczak, *Inorg. Chem.*, **16**, 3027 (1977).

[81] J. P. Sheckelton, J. R. Neilson, D. G. Soltan, and T. M. McQueen, *Nat. Mater.*, **11**, 493, (2012).

[82] J. P. Sheckelton, J. R. Neilson, and T. M. McQueen, *Materials Horizons*, **2**, 76 (2015).

[83] A. Kandala, A. Mezzacapo, M. Takita, M. Brink, J. M. Chow, and J. M. Gambetta, *Nature*, **549**, 242 2017.

[84] T. Lanting *et al.*, *Phys. Rev. X.*, **4**, 021041, (2014).

[85] Y. Tokura, M. Kawasaki, and N. Nagaosa, *Nat. Phys.*, **13**, 1056 (2017).

[86] Z. Chu, M. Javad, P. Asl, and S. Dong, *J. Phys. D: Appl. Phys.*, **51**, 243001, (2018).

[87] J. L. Schmehr, M. Aling, E. Zoghlin, and S. D. Wilson, arXiv:1902.05937 (2019).

[88] Eun-Ah Kim, MINT (2019).

Index

9 789811 219368

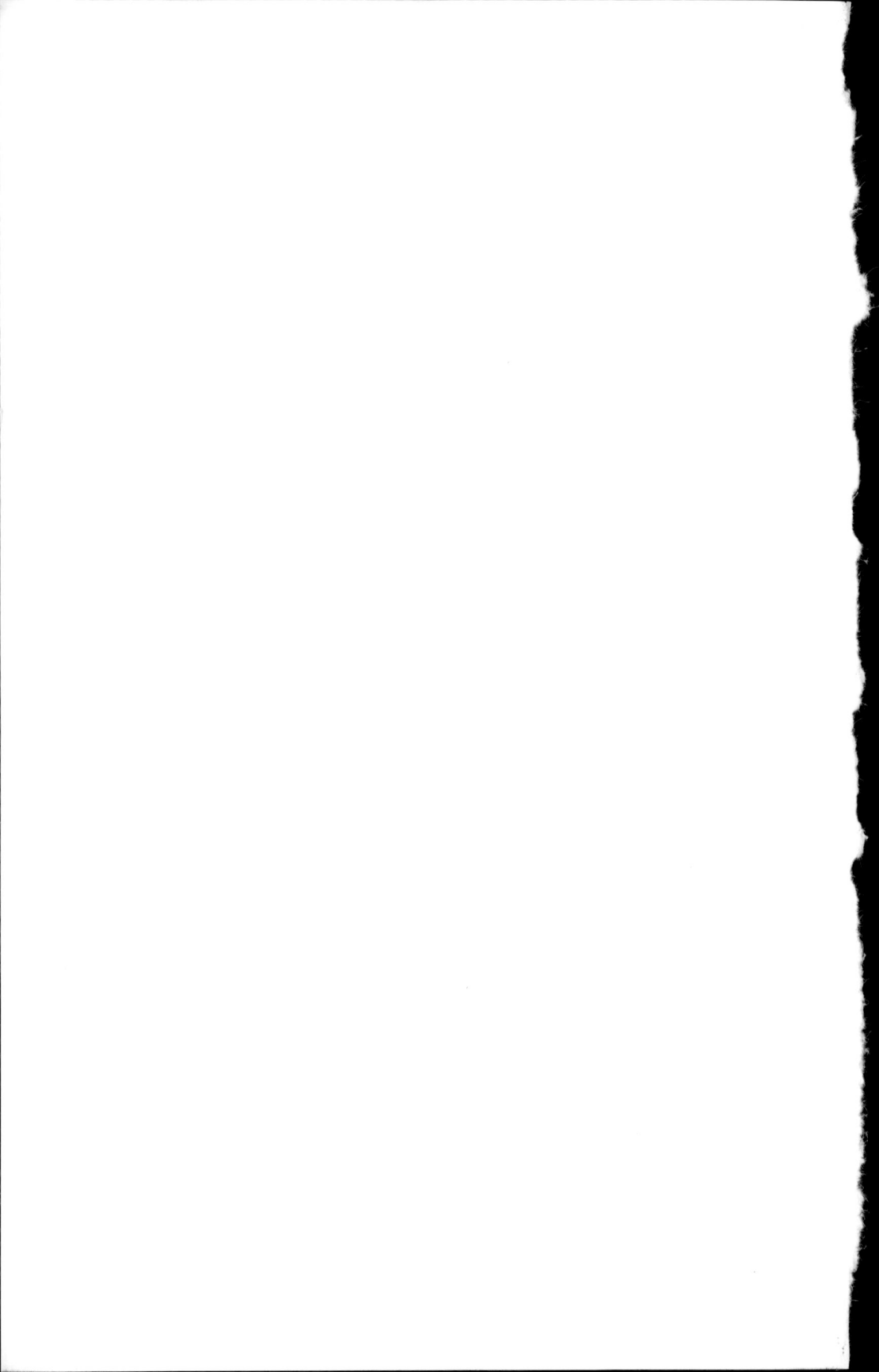